MINIMUM REINFORCEMENT IN CONCRETE MEMBERS

MINIMUM REINFORCEMENT IN CONCRETE MEMBERS

Editor: Alberto Carpinteri
Chair of Structural Mechanics
Politecnico di Torino
Torino, Italy

ESIS Publication 24

1999

Elsevier

Amsterdam • Lausanne • New York • Oxford • Shannon • Singapore • Tokyo

ELSEVIER SCIENCE Ltd
The Boulevard, Langford Lane
Kidlington, Oxford OX5 1GB, UK

First edition 1999

British Library Cataloguing in Publication Data
A catalogue record for this book is available from the British Library

Library of Congress Cataloging in Publication Data
```
Minimum reinforcement in concrete members / edited by Alberto
  Carpinteri. -- 1st ed.
     p.   cm. -- (ESIS publication ; 24)
   ISBN 0-08-043022-8
   1. Concrete beams--Testing. 2. Flexure.  I. Carpinteri, A.
  II. Series.
  TA683.5.B3M56  1999
  624.1'83423--dc21                              99-18029
                                                 CIP
```

ISBN: 0 08 043022 8

♾ The paper used in this publication meets the requirements of ANSI/NISO Z39.48-1992 (Permanence of Paper).

Printed and bound in the United Kingdom
Transferred to Digital Printing, 2011

Contents

Other titles in the ESIS Series

For information on how to order titles 1–23, please contact MEP Ltd, Northgate Avenue, Bury St Edmonds, Suffolk, IP32 6BW, UK

Preface

The ESIS-Technical Committee 9 on Concrete was established in 1990 and has met seven times (Noordwijk 1991, Vienna 1992, Brisbane 1993, Torino 1994, Zürich 1995, Poitiers 1996, Torino 1997). A round robin on "Scale effects and transitional failure phenomena of reinforced concrete beams in flexure" was proposed to European and extra-European laboratories and the following ones answered positively: Universidad Politecnica de Madrid (Spain), University of Sydney (Australia), Universität Stuttgart (Germany), Universitet Aalborg (Denmark), Università di Parma (Italy), Politecnico di Torino (Italy).

The central topic discussed in the committee is that of the *minimum reinforcement in concrete members*. The minimum amount of reinforcement is defined as that for which "peak load at first concrete cracking" and "ultimate load after steel yielding" are equal. In this way, any brittle behaviour is avoided as well as any localized failure, if the member is not over-reinforced. In other words, there is a reinforcement percentage range, depending on the size-scale, within which the plastic limit analysis may be applied with its static and kinematic theorems.

If we assume, as is usual in the codes of practice, a reinforcement amount proportional to the beam height h, the plastic bending moment, M_P, is proportional to the square of the beam height: $M_P \approx h^2$. On the other hand, if the reinforcement is assumed to react rigidly up to the crack propagation and to flow plastically only afterwards, the bending moment of crack propagation, M_C, is proportional to the beam height rised to 3/2: $M_C \approx h^{3/2}$. As a matter of fact, if steel-cover thickness and initial crack depth are assumed to be proportional to the beam height, e.g. both equal to $0.1\,h$, the stress-intensity factor K_I is proportional to the nominal stress and to the square root of beam height : $K_I \approx \sigma\sqrt{h} \approx \left(M/h^2 \right)\sqrt{h}$. Therefore, at the critical condition, when $K_I = K_{IC}$ we have $M_C \approx K_{IC}\,h^{3/2}$.

Since $M_P \approx h^2$ is a quantity of higher rank with respect to $M_C \approx h^{3/2}$, for smaller sizes we have $M_P < M_C$ and, for larger sizes, $M_P > M_C$. This means that, with the usual criteria, small beams tend to be under-reinforced, as well as large beams tend to be over-reinforced.

Carpinteri, Ferro, Bosco and El-Katieb propose a LEFM model, according to which reinforcement reactions are applied directly on the crack surfaces and a compatibility condition is locally imposed on the crack opening displacement in correspondence with the reinforcement. The theoretical model is found to provide a satisfactory estimate of the minimum percentage of reinforcement that depends on the scale and enables the element in flexure to prevent brittle failure. While the minimum steel percentage provided by Eurocode 2 and ACI are independent of the beam depth, the relationship established by the *brittleness number* N_P calls for decreasing values with increasing beam depths.

Lange-Kornbak and Karihaloo compare experimental observations with approximate nonlinear fracture mechanics predictions of the ultimate capacity of three-point bend, singly-reinforced concrete beams without shear reinforcement. The previous model, based on a zero crack opening condition and a fracture toughness accounting for slow crack growth, appears to be in good agreement with the observed failure mechanisms, although the test results indicate that a non-zero crack opening condition would improve the prediction, especially for lightly reinforced beams.

Ruiz, Elices and Planas introduce the so-called effective slip-length model, where the concrete fracture is described as a cohesive crack and the effect of reinforcement bond-slip is incorporated. Although the beams considered are of reduced size, the properties of the microconcrete were selected so that the behaviour observed is representative of beams of ordinary size made of ordinary concrete. The model is able to capture even experimental details and to describe the transitional behaviour from brittleness to ductility. Based on numerical analyses, a closed-form expression is given for minimum reinforcement in bending. The comparison with building codes and formulas from other Authors shows that the proposed expression provides safer or cheaper reinforcement, respectively for smaller or larger beams.

Fantilli, Ferretti, Iori and Vallini analyze the transition from pre-cracked to post-cracked stage by assuming a bond-slip relationship and a cohesive crack. The minimum reinforcement ratio is computed and its size-effects are enlightened. In addition to the dimensional analysis reasons previously considered, even the multifractal scaling of tensile strength and fracture energy are taken into account.

Brincker, Henriksen, Christensen and Heshe performed a very extensive experimental activity. Tensile failure of concrete at early stages and compression failure of concrete at final stages are modelled using fracture mechanics concepts. In this way, the experimental size effects are properly simulated and explained.

Ozbolt and Bruckner carry out numerical analyses of reinforced concrete beams of different sizes, using a finite element code which is based on the nonlocal microplane model. They take into account even dynamic regimes after cracking, so that the minimum reinforcement ratio, after reaching a critical beam size, increases with increasing size. This deduction is not in agreement with the preceding ones and, therefore, needs further theoretical and experimental studies.

The Chairman of ESIS-TC9 on Concrete is grateful to the European Structural Integrity Society for the confidence granted to him, and, in particular, to the Presidents subsequently in charge during the last recent years: Dr. L.H. Larsson, Dr. I. Milne, Prof. D. François; and to the Secretary Prof. A. Bakker. Special thanks are also due to Prof. K.J. Miller, Editor-in-Chief of the FFEMS Journal, for his continuous and patient encouragement.

The Editor of the volume then wishes to acknowledge all the Authors of the chapters, for their qualified contributions, and, in particular, his friends and colleagues Prof. Manuel Elices and Prof. Bhushan Karihaloo, who from the very beginning have always believed in the present project.

December 1998   Alberto Carpinteri
                                                                Politecnico di Torino
                                                                Torino, Italy.

SCALE EFFECTS AND TRANSITIONAL FAILURE PHENOMENA OF REINFORCED CONCRETE BEAMS IN FLEXURE

A. CARPINTERI, G. FERRO, C. BOSCO and M. ELKATIEB

Department of Structural Engineering, Politecnico di Torino,
10129 Torino, Italy.

ABSTRACT

Experimental observations and numerical simulations are compared with theoretical results based on a LEFM model, according to which reinforcement reactions are applied directly on the crack surfaces and a compatibility condition is locally imposed on the crack opening displacement in correspondence with the reinforcement. The theoretical model is found to provide a satisfactory estimate of the minimum percentage of reinforcement that depends on the scale and enables the element in flexure to prevent brittle failure. While the minimum steel percentage provided by Eurocode 2 and ACI are independent of the beam depth, the relationship established by the brittleness number calls for decreasing values with increasing beam depths.

KEYWORDS

Minimum reinforcement, scale effects, singly-reinforced concrete beams, transitional failure phenomena, brittleness number.

INTRODUCTION

Reinforced concrete beams undergo different failure mechanisms by varying steel percentage and/or beam slenderness and/or beam size-scale (in the last variation, the steel percentage being constant). The three fundamental collapse mechanisms are the following:

- nucleation and propagation of cracks at the edge in tension;

- compression and crushing at the edge in compression;

- formation of inclined shear cracks.

As regards tensile failures, the minimum amount of reinforcement can be determined through the concepts of fracture mechanics [1-5], while the maximum inelastic rotational capacity can be considered only when failure shifts to compressive sides. Both such quantities were found as subjected to remarkable size effects. An extensive experimental research was proposed by ESIS Technical Committee 9 on Concrete in order to obtain a rational and unified explanation for the transitions usually observed between the above mentioned collapse mechanisms.

The influence of size on the inelastic rotational capacity has not been completely clarified (and demonstrated) yet. In fact the experimental data available up to a few years ago, mostly obtained by load-controlled tests on reinforced concrete beams with high ductility bars, show a considerable scatter. On the other hand, some numerical evaluations, assuming strain localization in the compression zone, indicate that plastic rotation depends on the scale (i.e. beam depth) and the experimental tests recently carried out seem to validate this dependence [6, 7].

The attention is focused onto the transition between failure mechanisms (from reinforcement failure to concrete crushing) and onto the minimum percentage of reinforcement that depends on the scale [8] and enables the element in flexure to prevent brittle failure.

First of all, a presentation of the theoretical model based on LEFM is reported, in which the reinforcement reactions are applied directly to the crack surfaces and the compatibility condition is locally imposed to the crack opening displacement in correspondence with the reinforcement. Such a theoretical approach appears to be very useful for estimating the minimum amount of reinforcement for members in flexure, assuming simultaneous concrete cracking and steel yielding (transitional condition).

In the second part of the chapter, the experimental results of three point bending tests performed on 45 reinforced concrete beams at the Department of Structural Engineering of the Politecnico of Torino, are presented. From these results, a confirmation of the empirical formula for the critical value of the brittleness number N_P can be evidenced.

In addition, a numerical simulation of the tests was performed. The results presented were obtained using a specific FE program [9], particularly aimed at the object of the present investigation.

THEORETICAL MODEL

Superimposed Effect of Multiple Loads on the Deformation of a Cracked Beam Element

In order to analyse the behaviour of a structure containing a cracked member it is necessary to know the relation between the load and the deformation of the member. When a cracked member undergoes multiple loads simultaneously and it can be considered to behave elastically, a superimposed effect on the deformation should be evaluated.

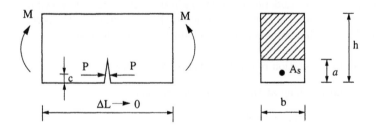

Figure 1: Cracked element

Let us consider the cracked member shown in Fig. 1, which undergoes simultaneously the bending moments M and the closing forces P applied onto the crack surfaces. It is possible to evaluate the angular deformation $\Delta\phi_{MP}$ produced by the forces P, together with the crack opening displacement $\Delta\delta_{PP}$, at the point where the forces P are applied, and, at the same time, the crack opening displacement $\Delta\delta_{PM}$ caused by the bending moment M, together with the angular deformation $\Delta\phi_{MM}$.

By linear superposition it is possible to write:

$$\Delta\delta = \Delta\delta_{PM} + \Delta\delta_{PP} = \lambda_{PM}M - \lambda_{PP}P \tag{1}$$

$$\Delta\phi = \Delta\phi_{MM} + \Delta\phi_{MP} = \lambda_{MM}M - \lambda_{MP}P \tag{2}$$

where λ_{MM}, λ_{PM}, λ_{MP}, λ_{PP}, are the compliances of the member due to the existence of the crack. The factors λ can be derived from energy methods considering the moments M acting simultaneously with the forces P.

If $\mathcal{G}$ and E are, respectively, the strain energy release rate and the Young's modulus of the material (the Poisson ratio ν is considered negligible), it follows that the variation ΔW of the total potential energy, is given by:

$$\Delta W = \int_0^c \mathcal{G}_M b\,\mathrm{d}x + \int_c^a \mathcal{G}_{(M+P)}b\,\mathrm{d}x = \int_0^c \frac{K_{IM}^2}{E}b\,\mathrm{d}x + \int_c^a \frac{(K_{IM}+K_{IP})^2}{E}b\,\mathrm{d}x =$$

$$= \int_0^c \frac{K_{IM}^2}{E}b\,\mathrm{d}x + \int_c^a \frac{K_{IM}^2}{E}b\,\mathrm{d}x + \int_c^a \frac{K_{IP}^2}{E}b\,\mathrm{d}x + 2\int_c^a \frac{K_{IM}K_{IP}}{E}b\,\mathrm{d}x = \tag{3}$$

$$= \int_0^a \frac{K_{IM}^2}{E}b\,\mathrm{d}x + \int_c^a \frac{K_{IP}^2}{E}b\,\mathrm{d}x + 2\int_c^a \frac{K_{IM}K_{IP}}{E}b\,\mathrm{d}x$$

where K_{IM} and K_{IP} are the stress-intensity factors due to bending moment M and to forces P, respectively.

Using Clapeyron's Theorem, it is possible also to write:

$$\Delta W = \frac{1}{2}M\Delta\phi_{MM} + \frac{1}{2}P\Delta\delta_{PP} + \frac{1}{2}(P\Delta\delta_{PM} + M\Delta\phi_{MP}). \tag{4}$$

Recalling that Betti's Theorem provides $P\Delta\delta_{PM} = M\Delta\phi_{MP}$, from eq.(3) and eq.(4) it is possible to obtain:

$$\frac{1}{2}M\Delta\phi_{MM} = \int_0^a \frac{K_{IM}^2}{E}b\,\mathrm{d}x \tag{5}$$

$$\frac{1}{2}P\Delta\delta_{PP} = \int_c^a \frac{K_{IP}^2}{E}b\,\mathrm{d}x \tag{6}$$

$$P\Delta\delta_{PM} = M\Delta\phi_{MP} = 2\int_c^a \frac{K_{IM}K_{IP}}{E}b\,\mathrm{d}x. \tag{7}$$

The stress-intensity factor produced at the crack tip by the moment M, can be expressed as [10-11]:

$$K_{IM} = \frac{M}{h^{3/2}b} Y_M(\xi) \tag{8}$$

with $\xi = a/h$, while the stress-intensity factor produced by the applied forces P acting at the level of reinforcement, i.e. at the distance c from the lower edge of the beam, is equal to [12]:

$$K_{IP} = \frac{2P/b}{\sqrt{\pi a}} F(\frac{c}{a}, \xi). \tag{9}$$

Rearranging the above expression, it is possible to write:

$$K_{IP} = \frac{P}{h^{1/2}b} Y_P(\frac{c}{h}, \xi) \tag{10}$$

where

$$Y_P(\frac{c}{h}, \xi) = F(\frac{c}{a}, \xi) \frac{2}{\sqrt{\pi \xi}}. \tag{11}$$

Function $Y_M(\xi)$ in eq.(8) is given by [10-11]:

$$Y_M(\xi) = 6 \times (1.99\xi^{1/2} - 2.47\xi^{3/2} + 12.97\xi^{5/2} - 23.17\xi^{7/2} + 24.80\xi^{9/2}) \tag{12}$$

for $\xi = \frac{a}{h} \leq 0.7$, while, function $F(\frac{c}{a}, \xi)$ in eq.(9) is given by [12]:

$$F(\frac{c}{a}, \xi) = \frac{3.52(1 - c/a)}{(1-\xi)^{3/2}} - \frac{4.35 - 5.28c/a}{(1-\xi)^{3/2}} + \left[\frac{1.30 - 0.30(c/a)^{3/2}}{(1-(c/a)^2)^{-1/2}} + 0.83 - 1.76c/a \right] [1 - (1 - c/a)] \tag{13}$$

for $a/h < 1$, $c/a < 1$.

Substituting eq.(8) into eq.(5) and dividing by M^2, the compliance λ_{MM} (rotation produced by $M = 1$), can be expressed as [13]:

$$\lambda_{MM} = \frac{2}{h^2 bE} \int_0^\xi Y_M^2(\xi) \mathrm{d}\xi. \tag{14}$$

In the same way, from eq.(10) and eq.(6), the compliance λ_{PP} (crack opening displacement produced by $P = 1$) becomes:

$$\lambda_{PP} = \frac{2}{bE} \int_{c/h}^\xi Y_P^2(c/h, \xi) \mathrm{d}\xi. \tag{15}$$

Eventually, substituting eq.(8) and eq.(10) into eq.(7) and dividing by the product PM, the compliance λ_{PM} (crack opening displacement produced by $M = 1$), or the compliance λ_{MP} (rotation produced by $P = 1$), can be expressed in the form:

$$\lambda_{PM} = \lambda_{MP} = \frac{2}{hbE} \int_{c/h}^\xi Y_M(\xi) Y_P(c/h, \xi) \mathrm{d}\xi. \tag{16}$$

Statically Indeterminate Reaction of Reinforcement

Let the cracked concrete beam element in Fig. 1 be subjected to the bending moment M, while the reinforcement transmits to the adjacent matrix surfaces an axial force, statically undetermined, equal to:

$$P = \sigma_s A_s \tag{17}$$

where A_s is the reinforcement area and σ_s the related stress.

If the displacement discontinuity in the cracked cross-section at the level of reinforcement is assumed to be zero, up to the moment of yielding or slippage of the reinforcement:

$$\Delta\delta = \Delta\delta_{PM} + \Delta\delta_{PP} = \lambda_{PM}M - \lambda_{PP}P = 0 \tag{18}$$

which is the displacement compatibility condition that allows to obtain the unknown force P as a function of the applied moment M. In fact, from eqs. (14-16) and considering eq.(1) and eq.(2), it follows that:

$$\frac{Ph}{M} = \frac{1}{r''(c/h, \xi)} \tag{19}$$

where:

$$r''(c/h, \xi) = \frac{\int_{c/h}^{\xi} Y_P^2(c/h, \xi)\mathrm{d}\xi}{\int_{c/h}^{\xi} Y_M(\xi)Y_P(c/h, \xi)\mathrm{d}\xi} = \frac{\lambda_{PP}}{\lambda_{PM}h}. \tag{20}$$

Considering a rigid-perfectly plastic behaviour of the reinforcement, the moment of plastic flow or slippage is obtained from eq.(19):

$$M_P = P_P h r''(c/h, \xi) \tag{21}$$

where $P_P = f_y A_s$ indicates the yielding (or pulling-out) force, achieved when $\sigma_s = f_y$ (yielding or pulling-out stress of reinforcement).

Stability of the Process of Concrete Fracture and Reinforcement Plastic Flow

The stress intensity factor at the crack tip is (eq.(8) and eq.(10)):

$$K_I = \frac{M}{h^{3/2}b}Y_M(\xi) - \frac{P}{h^{1/2}b}Y_P(c/h, \xi) \tag{22}$$

for $M < M_P$, or:

$$K_I = \frac{M}{h^{3/2}b}Y_M(\xi) - \frac{P_P}{h^{1/2}b}Y_P(c/h, \xi) \tag{23}$$

for $M > M_P$.

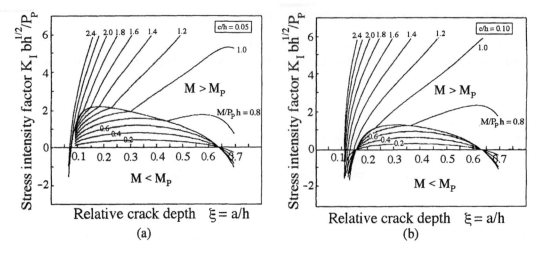

Figure 2: Dimensionless stress-intensity factor versus relative crack depth ξ varying the applied bending moment: (a) $c/h = 0.05$; (b) $c/h = 0.10$.

If $M < M_P$, eq.(19) can put into eq.(22):

$$K_I = \frac{M}{h^{3/2}b}Y_M(\xi) - \frac{1}{h^{1/2}b}Y_P(c/h,\xi)\frac{1}{r''(c/h,\xi)}\frac{M}{h} \tag{24}$$

and, in non-dimensional form:

$$\frac{K_I h^{1/2}b}{P_P} = \frac{M}{P_P h}\left[Y_M(\xi) - Y_P(c/h,\xi)\frac{1}{r''(c/h,\xi)}\right]. \tag{25}$$

In the same way, when $M > M_P$, eq.(23) becomes:

$$\frac{K_I h^{1/2}b}{P_P} = Y_M(\xi)\frac{M}{P_P h} - Y_P(c/h,\xi). \tag{26}$$

In order to plot the variations of the stress-intensity factor K_I against the crack depth ξ, it is necessary to verify the deformation condition for the reinforcement. This means that it is necessary to verify whether $\frac{M}{P_P h} \geq r''(c/h,\xi)$ or not. In the former case the reinforcement has yielded and it is possible to use eq.(26); otherwise eq.(25) must be used as stress σ_s is lower than f_y.

In Figs. 2.a and b the stress intensity factor K_I is reported for c/h=0.05 and c/h=0.10, respectively, against the crack depth ξ and varying the loading parameter $M/P_P h$. First of all, it is necessary to observe that the diagram is divided into two regions, characterized by different conditions of deformation for the reinforcement. The loading conditions and crack depths for which the assumed model predicts non-yielded reinforcement are represented below the separation line.

In Fig. 2.a all the curves characterized by values $M/P_P h \lesssim 0.8$ belong totally to the latter domain, while the curves characterized by values $M/P_P h \gtrsim 0.8$ belong only partially. This means that crack propagation is reached with an elastic condition for reinforcement in the entire range of ξ for $M/P_P h$ ratios lower than 0.8, while the same condition is verified only for limited crack lengths if $M/P_P h \gtrsim 0.8$. In other words, crack propagation is reached before steel yielding either for relatively high content of reinforcement or relatively limited crack depth.

The curves inside the domain where the reinforcement is in the elastic condition show a local or global maximum for $\xi \approx 0.35$. The locus of the maxima divides the zone where the cracking process is stable from the other zone where unstable propagation of the crack occurs. Beyond the value $M/P_P h \approx 1$ the curves do not present a maximum and this means that the cracking process is unstable for each crack depth $c/h < \xi < 0.7$.

The case $c/h = 0.10$ is represented in Fig. 2.b and the same trends are shown as in Fig. 2.a, whereas the elastic domain shrinks slightly.

Bending Moment of Concrete Fracture

Assuming that K_I is equal to the matrix fracture toughness K_{IC}, from eq.(22) it is possible to write:

$$\frac{M_F}{K_{IC} h^{3/2} b} = \frac{1}{Y_M(\xi)} + \frac{P}{K_{IC} h^{1/2} b} \frac{Y_P(c/h, \xi)}{Y_M(\xi)}. \tag{27}$$

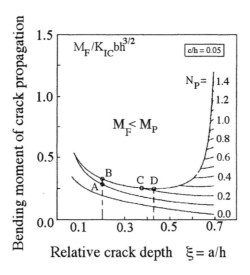

Figure 3: Dimensionless bending moment of crack propagation versus relative crack depth ξ varying the brittleness number N_P ($c/h=0.05$).

If the force P transmitted by the reinforcement is equal to $P_P = f_y A_s$ or, in other words, if the reinforcement yielding limit has been reached ($M = M_F \geq M_P$), eq.(27) becomes:

$$\frac{M_F}{K_{IC} h^{3/2} b} = \frac{1}{Y_M(\xi)} + N_P \frac{Y_P(c/h, \xi)}{Y_M(\xi)} \tag{28}$$

where the *brittleness number*

$$N_P = \frac{f_y h^{1/2}}{K_{IC}} \frac{A_s}{A} \qquad (29)$$

is introduced and $A = bh$ is the total cross-sectional area.

In the case $M = M_F < M_P$, i.e. when the reinforcement is in the elastic condition, we can consider the relation:

$$\frac{M_F}{K_{IC} h^{3/2} b} = \frac{1}{Y_M(\xi)} + N_P \frac{M_F}{M_P} \frac{Y_P(c/h, \xi)}{Y_M(\xi)} \qquad (30)$$

since, in that case, it is $\sigma_s/f_y = M_F/M_P$, and therefore $P = \sigma_s A_s = f_y(M_F/M_P)A_s$ in eq.(27).

Equation (30) may be modified by considering eq.(21) and eq.(29):

$$\frac{M_F}{K_{IC} h^{3/2} b} = \frac{1}{Y_M(\xi) - \dfrac{Y_P(c/h, \xi)}{r''(c/h, \xi)}}. \qquad (31)$$

Therefore, according to the model, when $M_F < M_P$ the moment of crack propagation M_F depends only on the relative crack depth ξ, and is not affected by the brittleness number N_P, i.e. it does not depend on the content of reinforcement but only on its relative position c/h.

The dimensionless fracture moment versus crack depth ξ, is reported in Fig. 3 for $c/h = 0.05$ and by varying N_P. The curves $N_P \lesssim 0.2$ are descending over the whole range ξ. This means that for low reinforced beams and/or for large cross-sections, the fracture bending moment decreases while the crack extends, i.e. an unstable fracture phenomenon occurs. For higher N_P values, the model predicts a stable fracture process with deep cracks. In particular this occurs for $N_P \gtrsim 0.3$.

A more attentive analysis of Fig. 3 can explain the behaviour of the cracked sections, varying loading conditions and initial crack length.

(a) Let N_P be equal to 0.1 and the initial relative crack depth $\xi=0.20$. When the applied bending moment M reaches M_F (point A in Fig. 3), the crack propagates and the phenomenon can not be stable anymore, the curve $N_P=0.1$ being always descending. Stable behaviour could only be obtained by reducing the external bending moment. It is worth noting that at point A the reinforcement has already yielded.

(b) Now let N_P be equal to 0.2 with the same initial relative crack depth previously assumed. When the bending moment of crack propagation is reached (point B in Fig. 3) the arc B-C of the transitional curve is followed. At point C the reinforcement, previously in elastic condition, yields, but since the curve is still descending, the fracturing process causes instability even in this case and develops up to complete failure of the cross-section.

(c) If N_P is greater than ≈ 0.25, crack propagation occurs until the transitional curve achieves its minimum, i.e. for $\xi=0.43$ (point D in Fig. 3). Beyond this point it is necessary to increase the bending moment to provoke a crack extension. Then, in these cases, the initial unstable phenomenon becomes stable.

(d) As a final example, let the initial relative crack depth ξ be grater than ≈ 0.43. In this case for any N_P greater than ≈ 0.25 the fracturing phenomenon is stable from the beginning. In fact the bending moment of crack propagation is reached in the ascending part of the curves, either in the transitional branch (when the elastic condition for reinforcement prevails) or, for deep cracks, in the plastic branch.

In conclusion, it is shown that the fracture process becomes stable only when N_P is sufficently high, i.e. when the cross-section of the beam is relatively small and/or the content of reinforcement relatively high, the crack extension being sufficiently deep.

Brittle or Ductile Collapse

For practical purposes it is interesting to determine when M_F (bending moment of crack propagation) is equal to M_P (bending moment of reinforcement yielding). This condition allows the determination of minimum steel reinforcement [4] which guarantees the stability of the beam behaviour, when the first cracking bending moment is reached.

When $M_F > M_P$, the elastic behaviour $(0 < M < M_P)$ is followed by a linear hardening behaviour $(M_P < M < M_F)$. The local rotation due to the applied loads is given by superimposed effects:

$$\Delta\phi = \lambda_{MM}M - \lambda_{MP}P \tag{32}$$

where λ_{MM} is obtained from eq.(14) and λ_{PM} from eq.(16). Then it is possible to obtain:

$$\Delta\phi = \lambda_{MM}(M - M_P), \quad \text{for } M > M_P. \tag{33}$$

Such a linear behaviour stops when crack propagation occurs. At that point, if the fracture phenomenon is unstable, function M versus $\Delta\phi$ presents a discontinuity and drops to a value obtainable considering the complete ligament disconnection. If the fracture phenomenon is stable, the discontinuity disappears and a continuous hardening response is obtained [1, 3].

The transitional value N_{PC} can be determined from eq.(28) if we impose $M_F = P_P(h - c) \approx P_P h$, i.e. no discontinuity in the M versus $\Delta\phi$ diagram.

In non-dimensional form:

$$\frac{M_F}{K_{IC}h^{3/2}b} = \frac{1}{Y_M(\xi)} + N_P\frac{Y_P(c/h,\xi)}{Y_M(\xi)} \approx \frac{P_P h}{K_{IC}h^{3/2}b} \tag{34}$$

and then, recalling eq.(29):

$$N_{PC} = \frac{1}{Y_M(\xi) - Y_P(c/h,\xi)}. \tag{35}$$

Thus it is possible, simply on the basis of the cross-sectional geometrical characteristics, to distinguish the cases of unstable fracture $(N_P < N_{PC})$ from those of stable fracture $(N_P > N_{PC})$.

EXPERIMENTAL DETAILS

Three point bending tests on 45 reinforced concrete beams had been planned at the Department of Structural Engineering of the Politecnico di Torino. Ten beams were cracked prior to loading due to the movement, so that only 35 beams have been really tested. Three values of slenderness L/h=6, 12, 18 (span to depth ratio) were taken into account. The beams were subdivided into three series, according to the different cross-sections: (A) 100x100 mm, (B) 100x200 mm and (C)

200x400 mm, while the effective depth to total depth ratio d/h was equal to 0.9 (Fig.4). The beams were reinforced only in tension with five different percentages: 0.12%, 0.25%, 0.50%, 1.00% and 2.00%. All the beams were cast from the same batch and no shear reinforcement in the central part of the beams was provided.

Table 1. Geometrical characteristics and steel reinforcement percentages of tested beams

Beam	Tension reinforcement	$A_s/(bh)$	N_P
A012-06	1 ϕ 5	0.20 %	0.187
A025-06	2 ϕ 5	0.39 %	0.374
A100-06	2 ϕ 8	1.00 %	1.019
A200-06	4 ϕ 8	2.00 %	2.038
A012-12	1 ϕ 5	0.20 %	0.187
A025-12	2 ϕ 5	0.39 %	0.374
A050-12	1 ϕ 8	0.50 %	0.510
A100-12	2 ϕ 8	1.00 %	1.019
A200-12	4 ϕ 8	2.00 %	2.038
A025-18	2 ϕ 5	0.39 %	0.374
A050-18	1 ϕ 8	0.50 %	0.510
A100-18	2 ϕ 8	1.00 %	1.019
A200-18	4 ϕ 8	2.00 %	2.381
B012-06	2 ϕ 5	0.20 %	0.265
B025-06	1 ϕ 8	0.25 %	0.360
B050-06	2 ϕ 8	0.50 %	0.721
B100-06	4 ϕ 8	1.00 %	1.441
B200-06	2 ϕ 16	2.00 %	2.322
B025-12	1 ϕ 8	0.25 %	0.360
B100-12	4 ϕ 8	1.00 %	1.441
B200-12	2 ϕ 16	2.00 %	2.322
C012-06	2 ϕ 8	0.12 %	0.255
C025-06	4 ϕ 8	0.25 %	0.510
C050-06	2 ϕ 16	0.50 %	0.821
C100-06	4 ϕ 16	1.00 %	1.642
C200-06	4 ϕ 20	2.00 %	2.810
C012-12	2 ϕ 8	0.12 %	0.255
C100-12	4 ϕ 16	1.00 %	1.642
C200-12	4 ϕ 20	2.00 %	2.810
C012-18	2 ϕ 8	0.12 %	0.255
C050-18	2 ϕ 16	0.50 %	0.821
C100-18	4 ϕ 16	1.00 %	1.642
C200-18	4 ϕ 20	2.00 %	2.810

A closed-loop servo-controlled testing machine was used. The tests were performed in displacement control for the beams with percentage of reinforcement larger than 0.50%, while for the others the crack mouth opening displacement (CMOD) control was used, in order to avoid sudden failure in the event of snap-back phenomena occurring. Top and bottom edge deformation were measured by means of potentiometric transducers with gauges of length equal to the depth of the beam, placed at 1/10 of the depth of the cross-section from both the beam extrados and intrados. The transducers had a 20 mm measuring range, while vertical displacements were measured by means of two transducers placed at midspan (Fig.5). Two additional transducers were used to measure the settlements at the supports, so that the real value of the midspan deflection was obtained by subtracting the average readings at the supports from the average readings at midspan. The end supports consisted in a fixed hinge and a roller enabling the beam to move horizontally, as shown in Fig.5. The load was transferred onto the beams by means of a platen having a length equal to half the beam depth and the same width in order to reduce stress concentration effects.

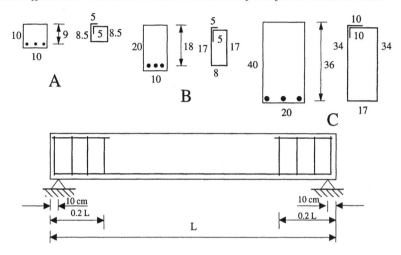

Figure 4: Arrangement of reinforcement.

The compressive strength of concrete was obtained from eight cubic specimens measuring 100 mm in side and the average value f_{cm} was 48.2 N/mm². The elastic modulus of concrete was determined from four specimens measuring 100x100x300 mm, which provided an average value equal to 35,000 N/mm². The fracture energy of concrete, determined according to RILEM recommendation [14] on six specimens, presented an average value $\mathcal{G}_F$ equal to 0.115 N/mm. The critical value of the stress-intensity factor can then be evaluated as:

$$K_{IC} = \sqrt{\mathcal{G}_F E} = 63.4 \ \text{N} \, \text{mm}^{-\frac{3}{2}}. \tag{36}$$

The steel bars had nominal diameters of 5, 8, 16 and 20 mm, respectively. Originally, the smallest bars should have a diameter of 4 mm; however from the post-mortem analysis on the beams the effective diameter came out to be equal to 5 mm. For this reason the percentages of reinforcement, reported in Table 1, are different from the prefixed ones. The 5 mm bars did not exhibit a well-defined yield point and the conventional yield limit, obtained from the stress-strain curve at 0.2% of permanent deformation, was equal to 604 N/mm². On the other hand, the yield strength for the bars of 8, 16 and 20 mm, equalled 643 N/mm², 518 N/mm² and 567 N/mm², respectively. The geometrical characteristics and steel reinforcement percentages of beams are reported in Table 1. The experimental setups and typical failure modes of some characteristic beams, are reported in Figs. 6-13.

EXPERIMENTAL RESULTS

In what follows, we shall consider the tests performed on 35 beams of classes A, B and C, with cross-sectional area equal to 100x100, 100x200 and 200x400 mm, respectively, and slenderness L/h equal to 6, 12 and 18. The values of the experimental parameters obtained from the considered beams are reported in Table 2.

For the beams of class A, with cross-sectional area equal to 100x100 mm, three diagrams are reported in Figs. 14-16. Each figure concerns one slenderness, and only the reinforcement ratio is varied. From the load versus deflection curves, it follows that the beam with reinforcement ratio 0.12% exhibits a brittle failure with the peak load higher than the reinforcement yielding load,

A. Carpinteri et al.

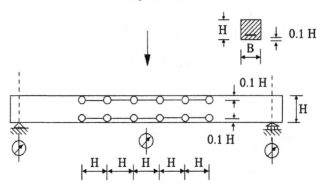

Figure 5: Setup of transducers.

whereas the beams with reinforcement ratio between 0.25% and 1.00% are characterized by a ductile response for all the three slendernesses. It is possible to observe in each diagram an initial linear elastic stage followed, in correspondence of the load of first cracking P_{cr}, by a more inclined line. The next stage is characterized by an horizontal line, typical of perfectly plastic behaviour. In this case, the compressive stresses are lower than the compressive strength of concrete and the reinforcement yields at the value f_y.

For the most reinforced beams (ρ=2.00%) and for L/h=6 and 12, a more brittle response is evident. Infact, after the peak load, the curves show an unstable response.

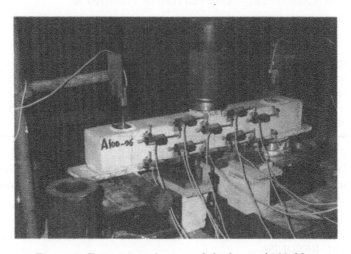

Figure 6: Experimental setup of the beam A100-06.

Figure 7: Experimental setup of the beam B100-12.

Figure 8: Experimental setup of the beam C100-06.

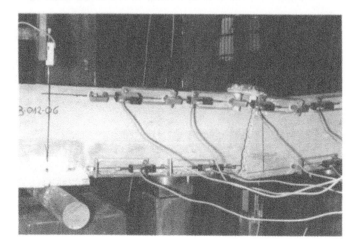

Figure 9: Brittle failure mode of the beam B012-06.

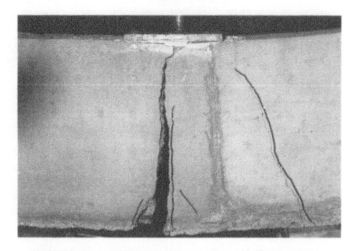

Figure 10: Brittle failure mode of the beam C100-06.

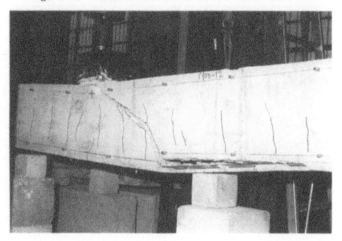

Figure 11: Shear failure mode of the beam C100-12.

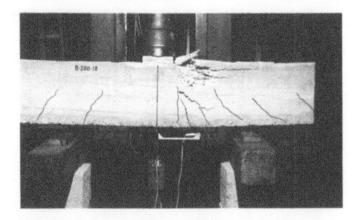

Figure 12: Brittle failure mode (crushing in compression) of the beam B200-12.

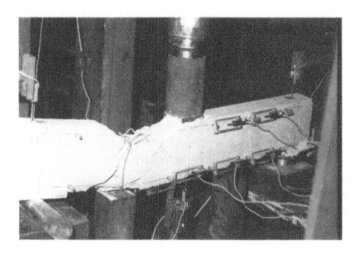

Figure 13: Brittle failure mode (crushing in compression) of the beam B200-06.

Table 2. Characteristic loads, deflections and rotations for the 35 experimental beams

Beam	Yielding load P_y (kN)	Peak load P_p (kN)	Ultimate load P_u (kN)	Midspan deflection at peak load δ_p (mm)	Plastic rotation at peak load θ_p	Ultimate rotation θ_u
A012-06	616	631	568	1.91	5.410 E-3	1.476 E-2
A025-06	1203	1232	1109	2.37	1.570 E-3	1.757 E-2
A100-06	3843	3910	3519	6.10	1.250 E-2	4.210 E-2
A200-06	6154	6276	5648	7.23	1.020 E-2	3.993 E-2
A012-12	304	321	289	2.96	4.100 E-4	1.004 E-2
A025-12	545	573	515	7.30	3.730 E-3	1.263 E-2
A050-12	923	945	850	10.18	7.300 E-3	2.769 E-2
A100-12	1662	1686	1518	13.11	1.175 E-2	4.060 E-2
A200-12	2753	2840	2556	11.00	2.600 E-3	1.692 E-2
A025-18	366	376	339	14.40	2.380 E-3	1.082 E-2
A050-18	628	642	578	22.60	2.860 E-3	4.400 E-3
A100-18	1141	1164	1048	25.40	1.113 E-2	2.445 E-2
A200-18	1936	1994	1795	16.80	1.070 E-3	3.700 E-3
B012-06	-	1450	1558	2.14	-	-
B025-06	2170	2295	2065	5.18	8.910 E-3	1.790 E-2
B050-06	3947	4185	3766	7.13	1.131 E-2	2.210 E-2
B100-06	7674	8300	7470	12.57	4.310 E-3	6.340 E-3
B200-06	9946	10753	9678	5.18	1.060 E-3	1.630 E-3
B025-12	1118	1134	1021	12.40	4.000 E-3	1.544 E-2
B100-12	4163	4236	3812	40.00	8.510 E-3	1.439 E-2
B200-12	6173	6223	5600	36.70	3.620 E-3	5.000 E-3
C012-06	4546	4665	4198	8.64	8.225 E-3	1.346 E-2
C025-06	9594	9869	8882	12.65	7.315 E-3	1.086 E-2
C050-06	14262	14527	13074	8.10	2.310 E-4	7.000 E-4
C100-06	22683	23803	21423	18.60	8.370 E-3	8.500 E-3
C200-06	30459	30954	27858	10.03	4.030 E-4	5.320 E-4
C012-12	1785	1803	1623	5.10	9.700 E-4	9.400 E-3
C100-12	10279	11403	10262	78.80	7.500 E-3	7.750 E-3
C200-12	16320	16632	14970	43.00	3.650 E-3	1.074 E-2
C012-18	802.3	834	751	24.00	3.600 E-3	1.315 E-2
C050-18	4007	4037	3633	49.20	3.670 E-3	4.000 E-3
C100-18	6328	6454	5809	131.60	9.700 E-3	1.650 E-2
C200-18	11348	11516	10364	69.74	9.000 E-4	1.023 E-2

In these cases a secondary crack crossing the flexural cracks leads to sudden brittle failure and the reinforcement does not reach the yield value, concrete failing in compression. Beam A200-18 shows an elastic-perfectly brittle response.

The experimental load versus deflection diagrams for the beams of class B, with cross sectional area equal to 100x200 mm, are reported in Figs. 17-18. As in the latter case, each diagram is related to a fixed slenderness, only the reiforcement ratio being varied. For slenderness 6, the curves show a ductile response for reinforcement ratios equal to 0.25 %, 0.50 % and 1.00 %. The beam B012-06 was cracked prior to loading, while the beam B200-06 (ρ=2.00%) shows a more brittle response with respect to the equivalent beam of class A. For slenderness 12 (ρ=0.12%, 1.00% and 2.00%) the increment in brittleness by increasing the reinforcement ratio is evident.

For the beams of class C, with a cross-sectional area equal to 200x400 mm, the diagrams are reported with the same sequence of the previous classes (Figs. 19-21). The beams with slenderness equal to 6 and low reinforcement ratio (ρ=0.12% and 0.25%) show a brittle response. For higher reinforcement ratios (ρ=0.50% and 1.00%), the response becomes ductile, whereas for the most reinforced beam (C200-06) an evident sudden and brittle failure occurs. For slenderness 12 (Fig. 20), the beam with ρ=0.12% shows a ductile response, while the beams with ρ=1.00% and ρ=2.00% show an unstable behaviour. In particular, the beam C200-12 shows a less brittle response than the beam B200-12, characterized by the same reinforcement ratio and slenderness. For slenderness 18 (Fig. 21), ductile behaviour is shown only by the beams with reinforcement ratios equal to 0.50% and 1.00%.

The behaviour of the beams with a low steel percentage (A012-12 and A012-06) presents peak loads exceeding the yielding loads. This phenomenon, formerly called *hyperstrength*, hard to explain from the theoretical point of view, has been justified by referring to a number of aspects which are usually neglected and yet play a significant role in structures with a low reinforcement percentage (as, for example, the nonlinearity of the tensile stress-strain relationship of concrete at the onset of the first cracking).

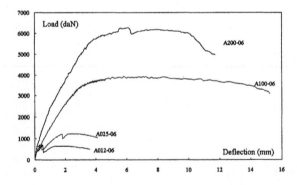

Figure 14: Load versus deflection diagrams of beams of class A and slenderness 6 by varying the reinforcement ratio.

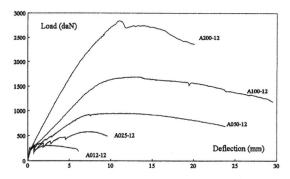

Figure 15: Load versus deflection diagrams of beams of class A and slenderness 12 by varying the reinforcement ratio.

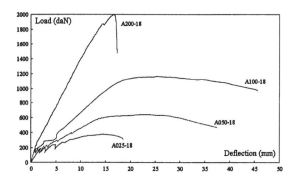

Figure 16: Load versus deflection diagrams of beams of class A and slenderness 18 by varying the reinforcement ratio.

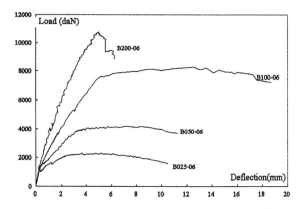

Figure 17: Load versus deflection diagrams of beams of class B and slenderness 6 by varying the reinforcement ratio.

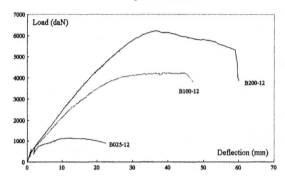

Figure 18: Load versus deflection diagrams of beams of class B and slenderness 12 by varying the reinforcement ratio.

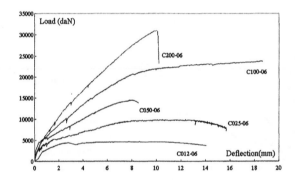

Figure 19: Load versus deflection diagrams of beams of class C and slenderness 6 by varying the reinforcement ratio.

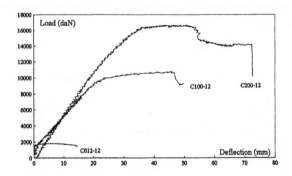

Figure 20: Load versus deflection diagrams of beams of class C and slenderness 12 by varying the reinforcement ratio.

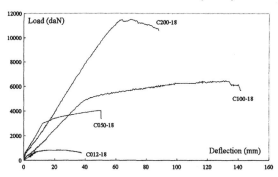

Figure 21: Load versus deflection diagrams of beams of class C and slenderness 18 by varying the reinforcement ratio.

NUMERICAL SIMULATIONS

A numerical investigation was planned in order to reproduce the experimental testing (even the potential response of the beams that have not been tested). In the simulations, performed with the FE code SBETA [9], the nonlinear softening law by Hordijk [15] was used. The program adopts the smeared crack model for modelling the cracks. The meshes used consisted in 160 to 500 elements, depending on the beam size. The number of mesh elements for each type of beam is reported in Table 3. The size of the element was chosen so to maintain a constant ratio between the depth of the beam and the depth of the element.

In Figs. 22-36 the dimensionless load versus deflection curves are reported. The dimensionless load is obtained by dividing P by $K_{IC}h^{0.5}b$, while the dimensionless deflection is obtained by dividing δ by h. As for the experimental diagrams, each diagram is characterized by a beam type (fixed depth and slenderness), only the reinforcement ratio being varied.

A typical ductile failure is described in Fig. 37, while in Fig. 38 the brittle failure (shear failure) is related to the beam C200-06. In Figs. 22-36, the dimensionless curves of numerical simulation are compared with the experimental ones. Each diagram represents the curves by varying the beam depth, slenderness and reinforcement ratio being constant. From the diagrams, it is possible to affirm that the numerical simulations approximate the experimental ones rather satisfactorily.

Table 3. Mesh used for the numerical simulations

Type of beam	Number of elements	Element size (cm)
A-06	200	2x2
A-12	350	2x2
A-18	500	2x2
B-06	170	4x4
B-12	320	4x4
B-18	470	4x4
C-06	160	8x8
C-12	310	8x8
C-18	460	8x8

Obviously, for the numerical simulations only one set of mechanical properties was used, although the real beams presented different reinforcement properties. The following mechanical parameters were used: f_c =35 N/mm^2; f_y =500 N/mm^2; E_c =30000 N/mm^2; E_s =200000 N/mm^2; ν =0.2; f_t =4 N/mm^2; $\mathcal{G}_F$ =0.1 N/mm.

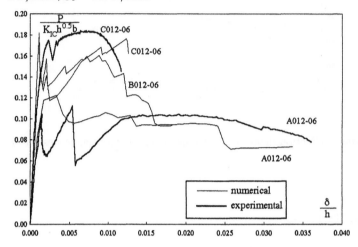

Figure 22: Dimensionless load versus deflection diagrams of beams of slenderness 6 and reinforcement ratio 0.12% by varying the beam depth.

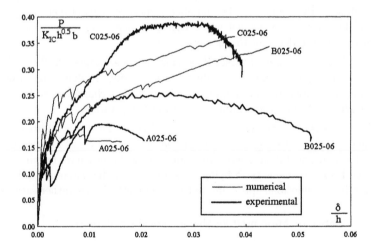

Figure 23: Dimensionless load versus deflection diagrams of beams of slenderness 6 and reinforcement ratio 0.25% by varying the beam depth.

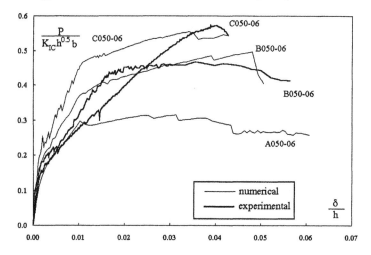

Figure 24: Dimensionless load versus deflection diagrams of beams of slenderness 6 and reinforcement ratio 0.50% by varying the beam depth.

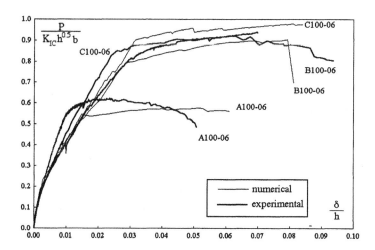

Figure 25: Dimensionless load versus deflection diagrams of beams of slenderness 6 and reinforcement ratio 1.00% by varying the beam depth.

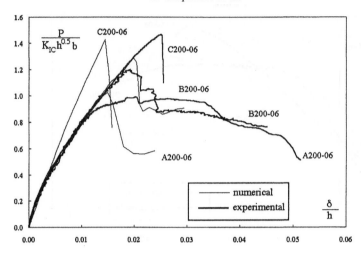

Figure 26: Dimensionless load versus deflection diagrams of beams of slenderness 6 and reinforcement ratio 2.00% by varying the beam depth.

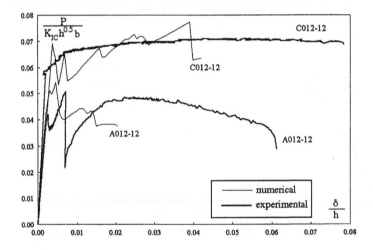

Figure 27: Dimensionless load versus deflection diagrams of beams of slenderness 12 and reinforcement ratio 0.12% by varying the beam depth.

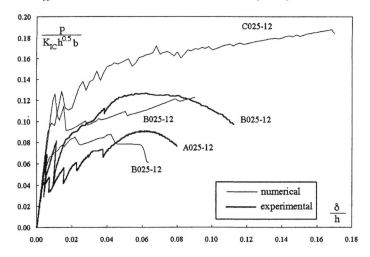

Figure 28: Dimensionless load versus deflection diagrams of beams of slenderness 12 and reinforcement ratio 0.25% by varying the beam depth.

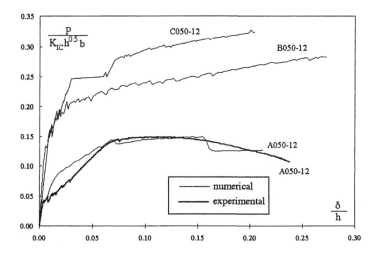

Figure 29: Dimensionless load versus deflection diagrams of beams of slenderness 12 and reinforcement ratio 0.50% by varying the beam depth.

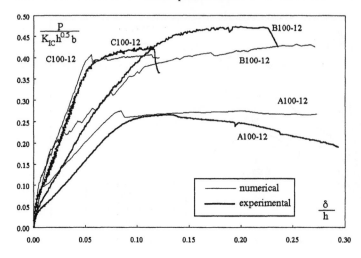

Figure 30: Dimensionless load versus deflection diagrams of beams of slenderness 12 and reinforcement ratio 1.00% by varying the beam depth.

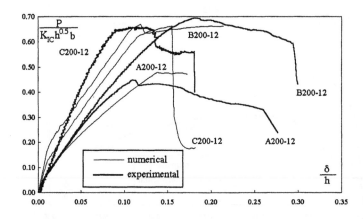

Figure 31: Dimensionless load versus deflection diagrams of beams of slenderness 12 and reinforcement ratio 2.00% by varying the beam depth.

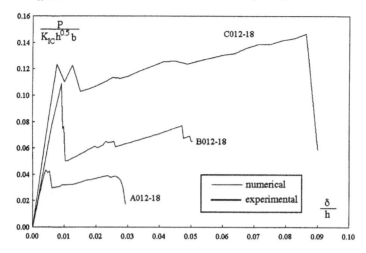

Figure 32: Dimensionless load versus deflection diagrams of beams of slenderness 18 and reinforcement ratio 0.12% by varying the beam depth.

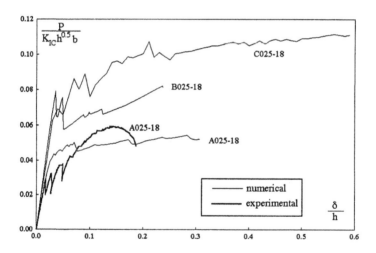

Figure 33: Dimensionless load versus deflection diagrams of beams of slenderness 18 and reinforcement ratio 0.25% by varying the beam depth.

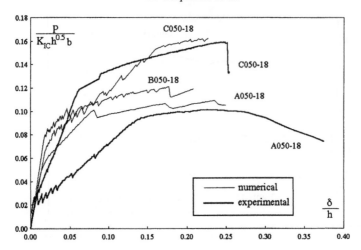

Figure 34: Dimensionless load versus deflection diagrams of beams of slenderness 18 and reinforcement ratio 0.50% by varying the beam depth.

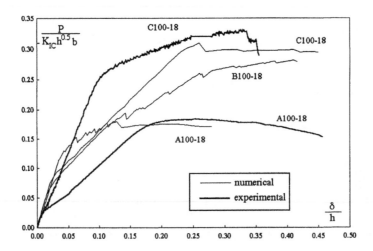

Figure 35: Dimensionless load versus deflection diagrams of beams of slenderness 18 and reinforcement ratio 1.00% by varying the beam depth.

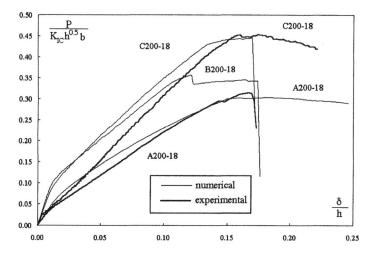

Figure 36: Dimensionless load versus deflection diagrams of beams of slenderness 18 and reinforcement ratio 2.00% by varying the beam depth.

Figure 37: Example of beam deformation with ductile response.

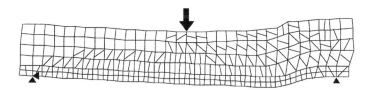

Figure 38: Example of beam deformation with compression failure.

MINIMUM REINFORCEMENT

Recently it has been proved experimentally that the minimum percentage of reinforcement that enables the element to prevent brittle failure, depends on the scale [1, 3]. With a classical approach these results should not be found nor predicable.

Minimum reinforcement provisions are not definitively settled in Standard Codes. The minimum percentage versus the beam depth is reported in Fig. 39 for four different concrete strengths. In the same diagram, the corresponding values obtained from EC2 [16] are reported, while the ACI 318 Code [17] provides a single percentage of steel.

A. Carpinteri et al.

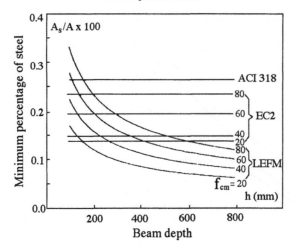

Figure 39: Minimum steel percentage versus beam depth.

The dependence of N_P from the beam depth causes the decreasing of the minimum steel percentage with increasing the beam depth, while the values supplied by the codes are independent of the beam depth.

The dimensionless load versus deflection diagrams are reported in Figs. 22-36. The dimensionless load is obtained by dividing P by $K_{IC}h^{0.5}b$, while the dimensionless deflection is obtained by dividing δ by h. The brittleness numbers $N_P = \frac{f_y h^{\frac{1}{2}} A_s}{K_{IC} A}$, A_s being the steel area and A the cross-sectional area, are reported for each beam in Table 1. Applying the formula proposed by Carpinteri and Bosco [5]:

$$N_{PC} = 0.1 + 0.0023\, f_c \qquad (37)$$

the transitional value $N_{PC}=0.211$ for the experimental tests is obtained.

From Table 1 it is possible to verify that only two beams (A012-06 and A012-12) are characterized by a brittleness number lower than the transitional one. In Fig. 22 (A012-06) and Fig. 27 (A012-12), it is shown how these two beams present a peak load higher than the yielding branch. The brittleness number appears to describe the ductile-brittle transition very satisfactorily. The same behaviour has been found in the numerical simulations (Fig. 22 and Fig. 27).

CONCLUSIONS

From the theoretical model with the forces transmitted by the reinforcement directly to the crack surfaces, it is possible to confirm the existence of different behaviours at failure, from ductile to brittle, the latter occurring when the beam depth overtakes certain values and/or the reinforcement content is sufficiently low. The model, as confirmed by the experimental results reported in this chapter and by other researchers [18], works very well for low reinforced concrete beams, the crack mouth opening displacement compatibility condition, taken into account to determine the force transmitted by the reinforcement, being very close to the real failure mechanism.

The value of the critical brittleness number N_{PC}, that separates brittle from ductile collapses, proposed by the two senior authors, is confirmed by the experimental as well as by the numerical results. Such a theoretical approach appears to be very useful for estimating the critical condition between stable and unstable crack propagation as well as the minimum reinforcement.

In conclusion, the utilization is proposed of a simple relationship, theoretically derived and experimentally and numerically confirmed, to calculate the minimum percentage of reinforcement for concrete members in flexure.

ACKNOWLEDGEMENTS

The present research was carried out with the financial support of the Ministry of University and Scientific Research (MURST) and the National Research Council (CNR).

REFERENCES

1. Carpinteri, A. (1981). "A fracture mechanics model for reinforced concrete collapse", IABSE Colloquium on Advanced Mechanics of Reinforced Concrete, Delft University of Technology, Delft, pp. 17-30.

2. Bosco, C. and Carpinteri, A. (1993) "Scale effects and transitional failure phenomena of reinforced concrete beams in flexure", ESIS Technical Committee 9, Round Robin Proposal, Department of Structural Engineering, Politecnico di Torino, Italy.

3. Carpinteri, A. (1984). "Stability of fracturing process in RC beams", *J. Struct. Eng. (ASCE)* **110**, 544-558.

4. Bosco, C., Carpinteri, A. and Debernardi, P.G. (1990). "Minimum reinforcement in high-strength concrete", *J. Struct. Eng. (ASCE)* **116**, 427- 437.

5. Bosco, C. and Carpinteri, A. (1992). "Softening and snap-through behavior of reinforced elements", *J. Eng. Mechanics (ASCE)* **118**, 1564-1577.

6. Bosco, C.,Carpinteri, C., and Debernardi, P.G. (1992). "Scale effect on plastic rotational capacity of r.c. beams", in *Fracture Mechanics of Concrete Structures*, Breckenridge, Z.P. Bažant (Ed.), Elsevier Applied Science, London, 1992, pp. 735-740.

7. Bosco, C. and Debernardi, P.G. (1992). "Experimental investigation on the ultimate rotational capacity of r.c. beams", Internal Report n.36, Dept. of Structural Engineering, Politecnico di Torino.

8. Bosco, C., Carpinteri, C., and Debernardi, P.G.(1990). "Fracture of reinforced concrete: scale effects and snap-back instability", *Eng. Fract. Mech.* **35**, 665-677.

9. SBETA Program Documentation (1992). Prague.

10. Okamura, H., Watanabe, K. and Takano, T. (1973). "Application of the compliance concepts in fracture mechanics", ASTM STP 536, pp. 423-438.

11. Okamura, H., Watanabe, K. and Takano, T. (1975). "Deformation and strength of cracked member under bending moment and axial force", *Eng. Fract. Mech.* **7**, pp. 531-539.

12. Tada, H., Paris, P. and Irwin, G. (1963). "The Stress Analysis of Cracks Handbook", Del Research Corporation, St. Louis, Missouri, Part 2, pp. 16-17.

13. Bosco, C. and Carpinteri, C. (1992). "Fracture mechanics evaluation of minimum reinforce-ment in concrete structures", in Application of Fracture Mechanics to Reinforced Concrete, A. Carpinteri (Ed.), Elsevier Applied Science, London, pp. 347-377.

14. RILEM Technical Committee 50 (1985). "Determination of the fracture energy of mortar and concrete by means of three-point bend tests on notched beams", Draft Recommendation, *Materials and Structures* **18**, pp. 287-290.

15. Hordijk, D. A. (1991). "Local approach to fatigue of concrete", Doctoral Thesis, Technische Universiteit Delft.

16. Comitée Euro-International du Béton (C.E.B.), Model Code 1990.

17. American Concrete Institute (1983). Building Code-Requirements for Reinforced Concrete (ACI 318-89), Detroit, Michigan.

18. Lange-Kornbak, D., Karihaloo, B.L. (1997). "Fracture mechanical calculation of flexural and diagonal tension strengths of singly-reinforced beams" ICF 9, Sydney, Australia.

FRACTURE MECHANICAL PREDICTION OF TRANSITIONAL FAILURE AND STRENGTH OF SINGLY-REINFORCED BEAMS

D. LANGE-KORNBAK

Danish Building Research Institute,
DK-2970 Hoersholm, Denmark

B. L. KARIHALOO

Civil Engineering Division, Cardiff University,
Cardiff CF2 3TB, Wales, UK

ABSTRACT

Experimental observations are compared with approximate nonlinear fracture mechanical predictions of the ultimate capacity of three-point bend, singly-reinforced concrete beams without shear reinforcement. The superposition model is found to provide a better estimate of the shear strength in diagonal tension failure than the size effect model. Moreover, Carpinteri's model based on a transitional brittleness number, a zero crack opening condition ($\delta = 0$) and a fracture toughness accounting for slow crack growth appears to be in good agreement with the observed failure mechanisms and ultimate (and residual) strength in the case of ductile (and brittle) flexural failure, although the test results indicate that a non-zero crack opening condition ($\delta \neq 0$) would improve the prediction, especially for lightly reinforced beams.

KEYWORDS

Approximate nonlinear fracture mechanics, superposition model, size effect model, singly-reinforced concrete beams, shear strength, flexural strength, diagonal tension failure, transitional brittleness number.

INTRODUCTION

A number of attempts have been made at predicting the strength of three-point bend, singly-reinforced concrete beams without shear reinforcement by means of fracture mechanics (for an overview, see [1]). Thus, the shear capacity due to diagonal tension failure has been established with the use of nonlinear fracture mechanics based on the fictitious crack model [2] and of an approximate, empirical method based on either the size effect model [3] or the two-parameter model [4,5]. While the former method requires numerical computations, the latter allows an analytical solution, thus making it particularly useful despite its approximate nature.

In addition, linear elastic fracture mechanics has been applied to the determination of the flexural strength [6,7]. An improvement to this procedure can be achieved by the use of approximate nonlinear fracture mechanics [8] or by considering the fracture process zone directly, assuming a stress distribution consistent with tension softening [9].

Table 1: Beam identification by specimen depth (A = 100 mm, B = 200 mm), reinforcement ratio ρ and slenderness ratio S/W

Beam dimensions					Reinforcement			Number	Beam
Length (mm)	Span (mm)	Width (mm)	Depth (mm)	S/W	Diameter (mm)	No. of bars	Percentage (%)	of specimens	identification code
800	600	100	100	6	4.0	1	0.125	3	A_012_06
					4.0	2	0.250	3	A_025_06
					7.5	2	0.880	3	A_088_06
2600	2400	100	200	12	4.0	2	0.125	2	B_012_12
					7.5	2	0.440	2	B_044_12
					16.0	2	2.010	2	B_200_12

Although these models have been available for some time now, no experimental studies have been undertaken to verify their accuracy. It is the aim of this paper to fill in this gap to a modest extent.

MATERIALS AND EXPERIMENTAL TECHNIQUE

Nine rectangular beams of size 600×100×100 mm and six of size 2400×100×200 mm (span S by width B by depth W) were tested as a part of the round-robin test program organised by ESIS Technical Committee 9. Three (two in the case of larger size) specimens of each size had the same arrangement of unanchored tension reinforcement (Table 1 and Fig. 1). The distance from the extreme tension fibre to the centroid of the steel bar was $c = 20$ mm $+ 0.5\, d_s$, with d_s the nominal diameter of bar.

The concrete mix was composed of Australian PC(A/MS/MA/G) cement (350 kg/m^3), tapwater (175 kg/m^3) and saturated, surface-dried aggregate (1825 kg/m^3) consisting of equal amounts of smooth, rounded river sand and partly smooth, partly rough, rounded river gravel with a maximum size (g) of 10 mm. This mix conformed to the round robin proposal [10] and an assumed density of 2350 kg/m^3.

At a maturity of 225 days (225 M-days, cf. [11]), the mix had the following mechanical properties [12]: uniaxial compressive strength $f'_c = 43.4$ MPa, modulus of elasticity $E = 21$ GPa (from 200×100×100 mm square prisms supplied with 0.127 mm teflon sheet at each end and loaded at a lateral deformation rate of 0.19 μm/s) and fracture toughness $K_{Ic} = 1.20$ MPa$\sqrt{\text{m}}$. These properties were required in subsequent predictions. The latter property, accounting for slow crack growth, was estimated by assuming mortar to be toughened by bridging, crack deflection and trapping and the concrete additionally by crack deflection, interfacial cracking and bridging (following the procedure outlined in [13]). Complete compressive stress-strain diagrams are available in [12]. For comparison, cylinders (of size 304.8×152.4 mm) without friction reducing material gave compressive strengths of 40.1 MPa at $9^{+}_{-}1$ M-days, 43.0 MPa at $27^{+}_{-}2$ M-days and 48.6 MPa $168^{+}_{-}1$ M-days in conjunction with a splitting tensile strength of 4.0 MPa at $168^{+}_{-}1$ M-days. These tests were stroke controlled at a rate of 5 μm/s (compression test) and 20 μm/s (splitting tensile test).

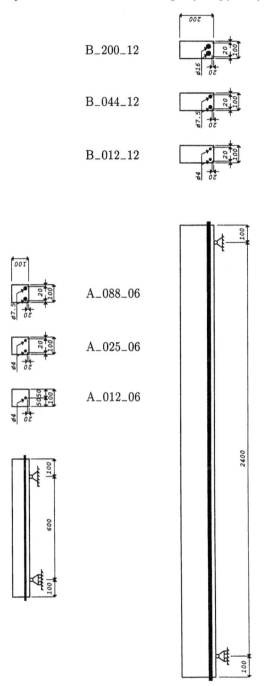

Figure 1: Arrangement of reinforcement.

Table 2: Mechanical properties of reinforcement

Nominal diameter (mm)	Yield stress (MPa)	Tensile strength (MPa)	Elongation at max load (%)	Elongation at failure (%)	Modulus of elasticity (GPa)
4	485	519	1.2	3.7	166
7.5	533	611	3.0	3.9	190
16.0	423	500	19.4	33.4	183

Table 3: Maturity and mass of beams at time of testing

Beam	Maturity at testing (M-days)	Mass (kg)
A_012_06 #1	163	19.2
#2	162	19.2
#3	106	19.3
A_025_06 #1	107	19.6
#2	163	19.6
#3	167	19.8
A_088_06 #1	110	19.4
#2	168	19.4
#3	167	19.6
B_012_12 #1	170	133.0
#2	170	129.0
B_044_12 #1	175	130.0
#2	171	131.0
B_200_12 #1	175	135.0
#2	176	134.5

The reinforcing steel was 4, 7.5 or 16 mm ribbed bars with the mechanical properties (the average of four tests) as reported in Table 2. The peak of the stress-strain curves yielded the tensile strength and elongation at maximum load, while for the two smallest bars the "yield stress", f_y, corresponded to a 0.2 % proof stress.

The beams were compacted with an immersion vibrator and stored at a dry temperature of 23 °C; two days in moulds under wet cloth and polyethylene sheets, then a further period in lime-saturated water (small beams) or under wet cloth and polyethylene sheets (large beams). The day before testing they were placed in the testing climate and levelled at the loading and support areas with a rubbing brick. They were then dried before being spray painted on one side. On the day of testing, a crack grid was marked on the painted surface and the beams were weighed (Table 3).

The testing machine, a DARTEC 80002 System I, was mounted with a 2000 kN servoactuator and load cell (sensitivity: 0.0005 V/kN) operating on the smallest built-in servovalve and a dynamic pump in order to maintain the load and control during control switching and fast load changes. The compression platens permitted rotation of an intermediate load transfer steel plate (dimensions and mass as given in Table 4). The average settlement of the supports, each taken as the mean

Table 4: Dimensions and mass of load transfer platens

Beam	Load transfer plate length × width × thickness (mm × mm × mm)	mass (kg)
A_ppp_s	50 x 100 x 50	4.297
B_ppp_s	100 x 100 x 50	6.383

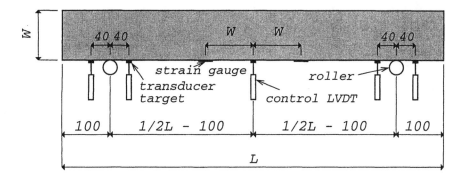

Figure 2: Setup of transducers.

value of two potentiometer readings on either side of the rollers, was subtracted from a LVDT reading at midspan to yield the true midspan deflection (Figure 2). The displacement transducers were centered on metallic targets (diameter 19 mm). The cracks were marked during breaks in the test. Some tests were controlled prior to peak by a half-Wheatstone bridge consisting of two strain gauges mounted at the bottom face symmetrically with respect to midspan at a distance equal to the beam depth (Figure 2). However, this type of control had a low success rate and midspan displacement control was employed instead for the remaining seven specimens, cf. Table 5.

The load-midspan deflection curves are given in Appendix A and summarized in Fig. 3. Note that the beam A_012_06 #2 was cracked prior to loading. (The jagged appearance of the curves is caused by breaks during which an extensometer was used to take vertical, lateral and oblique readings of target fields along the beams. These readings provided a means to determine the rotations along the beam. However, the load-strain curves for these fields are not presented in this work as the rotational capacity has been shown to be directly related to the rotation at the supports [14].) The rotations at the supports, $\alpha = \arctan((|d_1|+|d_2|)/80\text{mm})$ with d_i the deflection of the beam 40mm on either side of the support, are averaged and reported individually in cases where they differ significantly from each other. The rotations are plotted against the midspan deflection in Appendix B. In addition, the variation of the rotation with peak load (Table 6) is depicted in Figure 4.

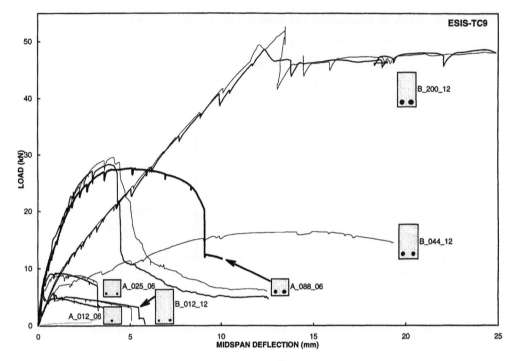

Figure 3: Summary of load-deflection curves.

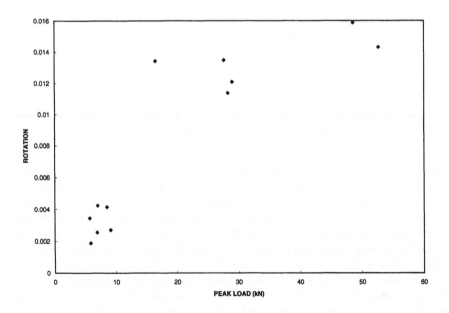

Figure 4: Relationship between peak load and corresponding rotation.

Table 5: Rates of strain, $\dot{\varepsilon}$, and midspan deflection, $\dot{\Delta}$, control signals. The prepeak rate refers to the rate between first cracking and peak loads. LC = control was lost

Beam	Control rate — Precrack			Prepeak		Postpeak	
	$\dot{\varepsilon}$ (μS/s)	$\dot{\Delta}$ (μm/s)	Δ range (mm)	$\dot{\varepsilon}$ (μS/s)	$\dot{\Delta}$ (μm/s)	Δ range (mm)	$\dot{\Delta}$ (μm/s)
A_012_06 #1	0.28			0.84			2.30
#2	0.28			0.84			2.30
#3		0.54			1.62		2.16
A_025_06 #1		0.54			1.62		2.16
#2	0.28			LC!			
#3	0.28			0.84		≤ 2.00	2.30
						> 2.00	3.50
A_088_06 #1		0.54			1.62		2.70
#2	0.28		≤ 0.70	0.84			6.00
			> 0.70	2.80			
#3	0.28		≤ 0.56	0.84		≤ 7.27	6.00
			0.56-0.72	1.68		> 7.27	9.95
			> 0.72	2.8			
B_012_12 #1		3.70			3.70	≤ 3.98	8.00
						> 3.98	10.00
#2		3.70			3.70		8.00
B_044_12 #1	0.35			LC!			
#2	0.35				3.70	≤ 15.81	7.50
						> 15.81	11.00
B_200_12 #1		5.00			15.0		45.00
#2		5.00			20.0	≤ 18.73	50.00
						> 18.73	90.00

The failure modes are presented in photographic form in Appendix C. The illuminated numbers in the bottom right of each photograph refer to the beam number, reinforcement ratio (%) and slenderness ratio. For example, 10 12 06 denotes beam number 1 with a reinforcement ratio $\rho = A_s/(BW) = 0.00125$ (with A_s the area of reinforcement) and a slenderness of 6. The beams with span 0.6 m and reinforcement ratio $\rho = 0.0088$ failed in diagonal tension except the beam A_025_06 #1 cured for only 107 days, instead of an average curing period of 169 days for the remaining beams. The beams with span 2.4 m and reinforcement ratio 0.02 failed initially in flexure followed by compressive failure of the concrete. All other beams failed in flexure.

DIAGONAL TENSION FAILURE

The shear strength P_u is determined by two approximate nonlinear fracture mechanical methods. First, the superposition method whereby P_u is regarded as being the sum of contributions from the reinforcement $(P_s)_u$ and concrete $(P_c)_u$, i.e. $P_u = (P_s)_u + (P_c)_u$. If x_{cr} and a_{cr} denote the location of the tip of diagonal tension crack at the instant of failure (Fig. 5), then following So and Karihaloo [5]

$$(P_s)_u = 2\frac{P_s(x_{cr})}{x_{cr}}\left(\frac{2}{3}W + \frac{1}{3}a_{cr} - c\right) \tag{1}$$

with

$$P_s(x_{cr}) = 0.5mF_1\pi d_s S F_2 \tau_{ult}\left(\frac{2x_{cr}}{S}\right)^N \le f_y A_s \tag{2}$$

Table 6: Peak load and corresponding rotation at supports. Values are not reported for tests where the control signal was lost (A_012_06 #2 was cracked prior to loading and is therefore ignored). Also shown are the failure modes

Beam	Peak Load (kN)	Rotation	Failure Mode
A_012_06 #1	6.87719	0.002558	Tension
#2			Tension
#3	6.98254	0.004245	Tension
A_025_06 #1	8.5194	0.004142	Tension
#2			Tension
#3	9.09862	0.002712	Tension
A_088_06 #1	27.668	0.013476	Tension
#2	28.9944	0.012095	Shear
#3	28.3086	0.011381	Shear
B_012_12 #1	5.62134	0.003443	Tension
#2	5.79198	0.001887	Tension
B_044_12 #1			Tension
#2	16.4776	0.013417	Tension
B_200_12 #1	52.6658	0.014292	Tension followed by compression
#2	48.5212	0.015871	Tension followed by compression

and

$$F_1 = \frac{93 + 135(0.5B/d_s - 1) - 7(0.5B/d_s - 1)^2}{93 + 135(B/d_s - 1) - 7(B/d_s - 1)^2}$$

$$F_2 = 0.3889 + 0.0592 \left(\frac{S}{W}\right) - 0.0017 \left(\frac{S}{W}\right)^2 \qquad (3)$$

$$\tau_{ult} = 0.4684\sqrt{f'_c} \left(\frac{\rho S}{2d_s}\right)^\gamma + 0.0271(c - 1.5d_s)$$

$\gamma = -0.8205d_s - 0.2933$, m is the number of bars and N is an exponent which depends on the bond-slip relationship. For unanchored bars and diagonal tension failure $N = 1.25$ seems appropriate [1]. Note that $a_{cr} = W$ corresponds to a brittle, sudden "diagonal tension failure", whereas $a_{cr} < W$ is the more ductile "shear-compression failure".

The contribution of concrete to the shear strength is obtained by an incremental procedure, proposed by Karihaloo [15]. This incremental procedure differs from the Jenq-Shah approach [4] in several ways. First, the contribution from the steel-concrete interaction is better approximated to reflect the full range of available test data. Secondly, the location of the incipient diagonal tension crack and the direction of its growth are varied with a view to delineating the critical diagonal tension crack. Thirdly, an allowance is made for the fact that the crack is eccentrically placed relative to the applied load, so that it is in a state of mixed mode fracture (modes I and II). Fourthly, the instant of growth and direction of propagation of the crack are determined by using a proper mixed mode fracture criterion. However, in the beams under consideration, the diagonal tension crack propagated almost in a straight line ($\theta = 24°$) and there was no compression failure

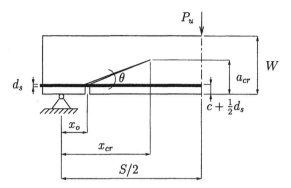

Figure 5: Crack configuration at ultimate load P_u for shear failure.

in concrete. Thus, the procedure converges in just one step, giving

$$(P_c)_u = K_{Ic} \left[\left(\frac{3Y_1(a_{cr}/W)x_{cr}\sqrt{a_{cr}}}{BW^2} \right)^2 + \left(\frac{Y_2(a_{cr}/W)\sqrt{a_{cr}}}{2BW} \right)^2 \right]^{-1/2} \qquad (4)$$

where the geometry functions are

$$(Y_1, Y_2) = \begin{cases} (2.722, 0), & a_{cr}/W = 0.55, 2S_1/S = 0 \\ (2.989, 2.141), & a_{cr}/W = 0.55, 2S_1/S = 1/6 \\ (3.113, 0), & a_{cr}/W = 0.60, 2S_1/S = 0 \\ (3.066, 2.902), & a_{cr}/W = 0.60, 2S_1/S = 1/6 \end{cases} \qquad (5)$$

with $S_1 = S/2 - x_{cr}$.

The second approximate method is based on the size effect model [3] and gives $P_u = 2V_u BW$, with

$$V_u = \frac{8\rho^{1/3}}{\sqrt{1 + W/(25g)}} \left[\sqrt{f_c'} + 3000 \sqrt{\rho/ \left(\frac{S}{2W} \right)^5} \right] \qquad (6)$$

in which f_c' and V_u are in psi.

Beams A_088_06 failed in diagonal tension and had $x_o = 20$ mm and $x_{cr} = 275$ mm. The size effect model (eqn 6) predicts a rather conservative value $P_u = 16.8$ kN. However, it has been pointed out that formula (6) is only applicable to low strength concretes ($f_c' < 30$ MPa) and that for concretes of higher strength the numerical factor 25 must be increased [16]. A numerical value of ∞ would still predict only $P_u = 19.2$ kN.

According to the superposition method used here, linear inter- and extrapolation in (5) gives $Y_1 = 4.9615$ and $Y_2 = 4.495$, whereby $P_u = (P_c)_u + (P_s)_u = 1.0$ kN $+ 23.2$ kN $= 24.2$ kN. This value is lower than the measured ultimate load of 28.65 kN, as the interlocking of the crack surfaces by coarse aggregate and the dowel action of the reinforcement have been neglected. Moreover, the vertical notch in the sub-problems represents only an approximation of the real inclined crack. A better approximation, cf. Fig. 6, may be to consider the case of a crack initiating at some angle θ_1 from the surface of a strip and subsequently kinking at an angle θ. The strip is subjected to a remote uniform tensile stress p acting parallel to the surface. Stress intensity factors for this

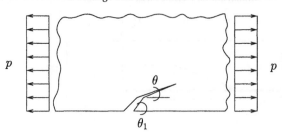

Figure 6: Alternative diagonal tension problem for the determination of mixed-mode stress intensity factors.

problem are tabulated in [17] for $\theta_1 = \pi/2$ and θ follows from $\tan\theta = (a_{cr} - c - 0.5d_s)/(x_{cr} - x_o)$. Assuming $p = 6M(x_{cr})/(BW^2)$, although a lower and linearly variable stress would be closer to reality, the model predicts $(P_c)_u = 2.3$ kN, so that $P_u = 25.5$ kN, which is in better agreement with the experimental value. This simple calculation clearly highlights the importance of considering the actual crack geometry and not its approximation by a vertical, albeit off-centre crack.

FLEXURAL FAILURE

Beams made from conventional, relatively brittle concrete typically fail in flexure by unstable crack propagation at bending moment M_F, preceding yielding (or slippage) of reinforcement at M_P. Bosco and Carpinteri [7] found that the ultimate bending moment $M_u = M_F$ for $N_p < N_{pc}$ and $M_u = M_P$ for $N_p > N_{pc}$, where the brittleness number $N_p = \rho f_y W^{1/2}/K_{Ic}$ and the transitional brittleness number $N_{pc} = 0.1 + 0.0023 f'_c$ (f'_c in MPa). (Note that if the concrete is relatively ductile, e.g. owing to fibre reinforcement, then yielding of the reinforcement is likely to precede unstable crack propagation.) Denoting by M the bending moment at the cracked cross section, then the crack with a mouth opening δ, is subjected to a closing force $P^* = P_s$, stemming from the reaction from the reinforcement, and a bending moment $M^* = M - P_s\left(\frac{W}{2} - c\right)$ due to the stress distribution in the uncracked ligament. Bosco and Carpinteri established that

$$M_P = \frac{0.5WBE\delta + f_y A_s W \int_\beta^{\alpha_P} Y_p^2(x,\beta)dx}{\int_\beta^{\alpha_P} Y_m(x)Y_p(x,\beta)dx} + f_y A_s\left(\frac{W}{2} - c\right) \tag{7}$$

$$M_F = K_{Ic}W^{3/2}B\left\{\frac{1}{Y_m(\alpha_F)} + N_p\frac{Y_p(\alpha_F,\beta)}{Y_m(\alpha_F)} + N_p\left(\frac{1}{2} - \frac{c}{W}\right)\right\} \tag{8}$$

where $(\beta = c/W) \leq x \leq (\alpha = a/W)$ and the geometry factors are [1]

$$Y_m(x) = 6\sqrt{x}\left(1.99 - 2.47x + 12.97x^2 - 23.17x^3 + 24.8x^4 + 60.5x^{16}\right) \tag{9}$$

$$Y_p(x,\beta) = \frac{2}{\sqrt{\pi x}}\left[\frac{3.52(1 - \beta/x)}{(1-x)^{3/2}} - \frac{(4.35 - 5.28\beta/x)}{(1-x)^{1/2}} + F_1(x,\beta)\right] \tag{10}$$

with

$$F_1(x,\beta) = \left\{\frac{1.3 - 0.3(\beta/x)^{3/2}}{\sqrt{1 - (\beta/x)^2}} + 0.83 - 1.76\beta/x\right\}\{1 - (1 - \beta/x)x\} \tag{11}$$

$Y_m(x)$ is valid for $0 \leq x \leq 0.8$, $S/W \geq 6$, and $Y_p(x,\beta)$ in the range $x < 1.0$, $\beta/x < 1.0$.

Table 7: Measured and predicted ultimate and residual loads

Beam	Brittleness no.	Crack/Beam depth		Ultimate load (kN)		Residual load (kN)	
		Ult. load	Res. load	Meas.	Predic.	Meas.	Predic.
A_012_06	0.16	0.35(-0.75)	0.80	6.93	6.48	4.62	3.59
A_025_06	0.32	0.75	-	8.81	7.01	-	-
B_012_12	0.23	0.30(-0.55)	0.70	5.71	7.02	5.07	4.33
B_044_12	0.88	0.60	-	16.48	16.36	-	-
B_200_12	3.16	0.55	-	50.59	56.45	-	-

The brittleness numbers N_p tabulated in Table 7 suggest that only beams A_012_06 with $N_p < N_{pc}$ = 0.20 failed in a brittle manner, i.e. $M_u = M_F$. However, the load-deflection curves reveal that this failure mode was also exhibited by beams B_012_12. This could be a result of using the approximate expression for $N_{pc} = 0.1 + 0.0023 f'_c$. The exact theoretical relation $N_{pc} = [Y_m(\alpha_F = \alpha_M) - Y_p(\alpha_F = \alpha_M, \beta)]^{-1}$ gives $N_{pc} = 0.23$, with $\alpha_F = a_F/W = 0.51$. The measurement of a_F at the initiation of unstable crack growth presents a difficult problem. The value indicated here is a reasonable estimate. The condition $N_p < N_{pc}$ therefore appears to hold also for beams B_012_12.

Assuming $\delta = 0$, Table 7 also summarizes the measured and predicted ultimate (i.e. maximum or peak) load, along with the residual (i.e. maximum post-peak) load corresponding to yielding of the steel for the beams exhibiting brittle failure. It is clear that a good agreement is obtained between measured and predicted values; the prediction, however, becomes less accurate as the amount of steel ρ is reduced. This may be attributed to the increase in crack opening δ with decreasing ρ which is not accounted for in the present calculations (cf. Eqn 7). For the predicted load at yielding of the reinforcement to be equal to the measured load, a crack opening $\delta = 0.162$mm, 0.168mm and 0.107mm is required for beams A_012_06, A_025_06, B_012_12, respectively. Formulae for the determination of δ (see e.g. Park and Paulay [18]) should be used with caution as they are usually designed to overestimate δ somewhat. For instance, the formula of Kaar and Hognestad or that of Gergely and Lutz gives values of δ which are 2.5 to 4 times larger than the above theoretical values leading to a significant overestimation of the load at the yielding of reinforcement.

The overestimation of the strength of beams B_200_12 can be attributed to the existence of a strain softening zone ahead of the crack tip, which reduces the geometry factors (9) and (10), thereby ultimately lowering M.

CONCLUSIONS

The flexural and diagonal tension failure loads of singly-reinforced beams without shear reinforcement obtained by testing the beams in three-point bending were compared with the predictions of approximate nonlinear fracture mechanical models. The superposition model was found to provide a better estimate of the shear strength in diagonal tension failure than the size effect model. The original superposition model was improved by considering the actual crack geometry and not its approximation by a vertical, albeit off-centre crack. Moreover, Carpinteri's model based on a zero crack opening condition ($\delta = 0$) and a fracture toughness accounting for slow crack growth gave results which were in good agreement with the measured ultimate (and residual) strength in the case of ductile (and brittle) flexural failure. The test results indicated that a non-zero crack opening condition ($\delta \neq 0$) would improve the prediction, especially for lightly reinforced beams. However, the crack opening would be significantly smaller than that predicted by traditional strength-based formulae.

REFERENCES

1. Karihaloo, B.L. (1995a). *Fracture Mechanics & Structural Concrete.* Longman Scientific & Technical, UK.

2. Gustafsson, P.J. and Hillerborg, A. (1988) *J. ACI* **85**, 286.

3. Bažant, Z.P. and Kim, J.K. (1984) *J. ACI* **81**, 456.

4. Jenq, Y.S. and Shah, S.P. (1989). In: *Fracture Mechanics: Application to Concrete*, V.C. Li and Z.P. Bazant (Eds). American Concrete Institute, Detroit, pp. 327-358. .

5. So, K.O. and Karihaloo, B.L. (1993) *J. ACI* **90**, 591.

6. Carpinteri, A. (1984) *ASCE J. Struct. Engng.* **110**, 544.

7. Bosco, C. and Carpinteri, A. (1992). In: *Applications of Fracture Mechanics to Reinforced Concrete*, A. Carpinteri (Ed.). Elsevier, London, pp. 347-377.

8. Karihaloo, B.L. (1992). In: *Applications of Fracture Mechanics to Reinforced Concrete*, A. Carpinteri (Ed.). Elsevier, London, pp. 523-546.

9. Baluch, M.H., Azad, A.K. and Ashmawi, W. (1992). In: *Applications of Fracture Mechanics to Reinforced Concrete*, A. Carpinteri (Ed.). Elsevier, London, pp. 413-436.

10. Bosco, C. and Carpinteri, A. (1993). ESIS Technical Committee 9 Round Robin Proposal, Dept. Struct. Engng., Politecnico di Torino, Italy.

11. Hansen, P.F. (1978). Hærdeteknologi 2: dekrementmetoden, BKF-Centralen (BKI), Aalborg Portland, Denmark, pp. 57-61 (in Danish).

12. Lange-Kornbak, D. and Karihaloo, B.L. (1994). Strain Softening of Concrete under Compression, Report to RILEM Committee 148SSC, Dept. Civil Engng., University of Sydney, Australia.

13. Lange-Kornbak, D. and Karihaloo, B.L. (1996) *Advanced Cem. Based Mater.* **3**, 124.

14. Ulfkjær, J.P. (1995). Presentation at the 5th meeting of ESIS-TC9 on Concrete, Zürich, Switzerland.

15. Karihaloo, B.L. (1995b). In: *Fracture Mechanics of Concrete Structures*, F.H. Wittmann (Ed.). AEDIFICATIO, Freiburg, pp. 1111-1123.

16. Bažant, Z.P. (1997). Private communication.

17. Nisitani, H. (1975) *Nihon Kikai Gakkai Rombunsio* **41**, 1103 (in Japanese) (reproduced in *Fracture Mechanics, Handbook*, Vol. 2, Nauka Dumka, Kiev, 1988, pp. 125-127 (in Russian)).

18. Park, R. and Paulay, T. (1975). *Reinforced Concrete Structures.* Wiley-Interscience, New York.

APPENDIX A

LOAD-MIDSPAN DEFLECTION CURVES

D. Lange-Kornbak and B. L. Karihaloo

A_012_06 #1

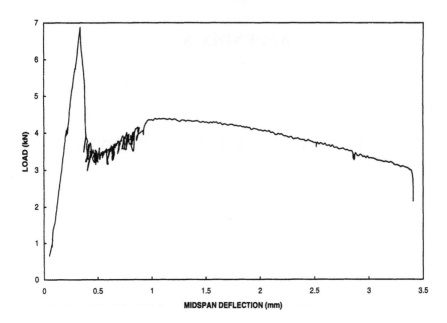

A_012_06 #2

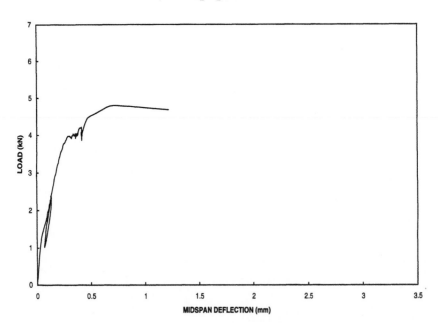

A_012_06 #3

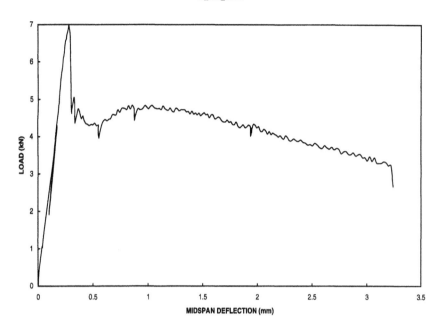

A_025_06 #1

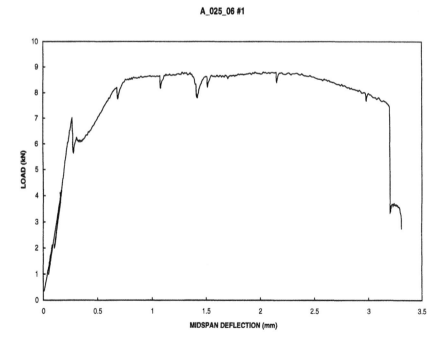

A_025_06 #3

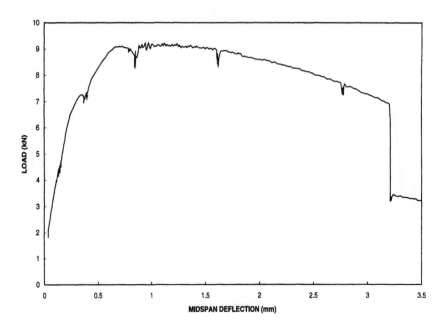

A_088_06 #1

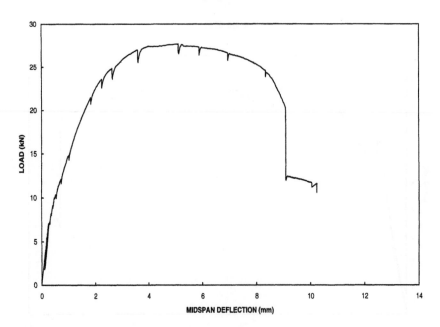

A_088_06 #2

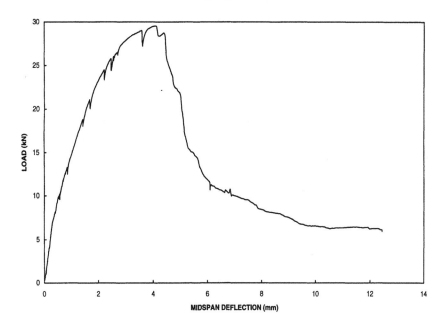

A_088_06 #3

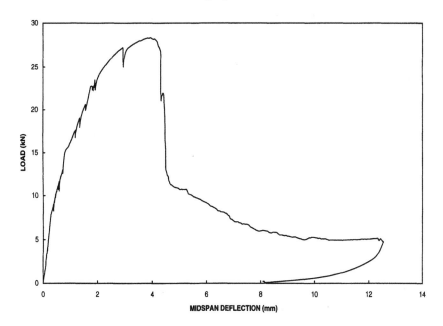

D. Lange-Kornbak and B. L. Karihaloo

B_012_12 #1

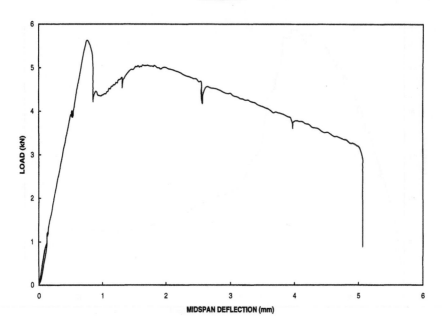

B_012_12 #2

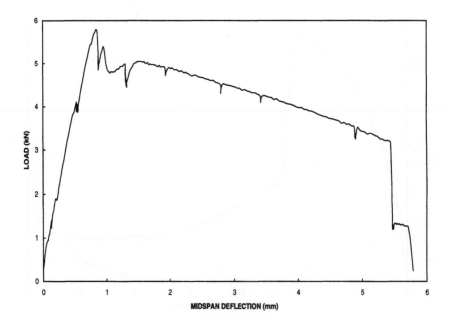

B_044_12 #2

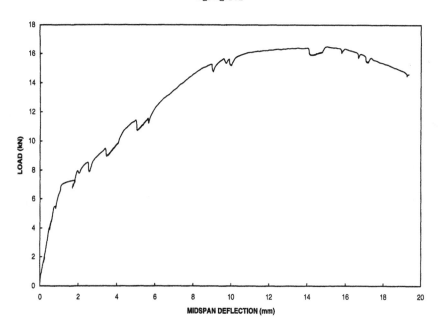

B_200_12 #1

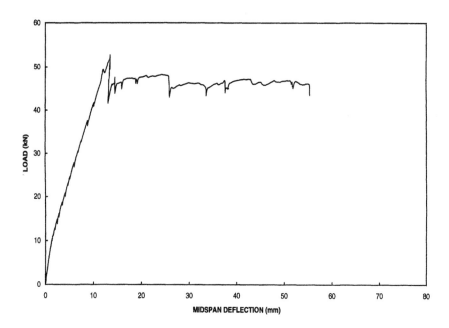

D. Lange-Kornbak and B. L. Karihaloo

B_200_12 #2

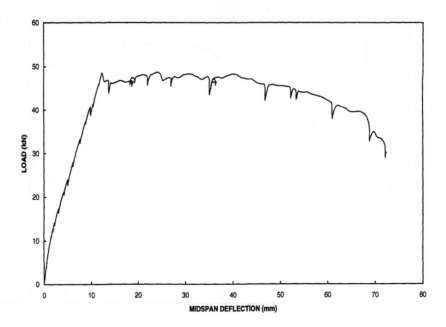

APPENDIX B

ROTATION AT SUPPORTS

D. Lange-Kornbak and B. L. Karihaloo

A_012_06#1

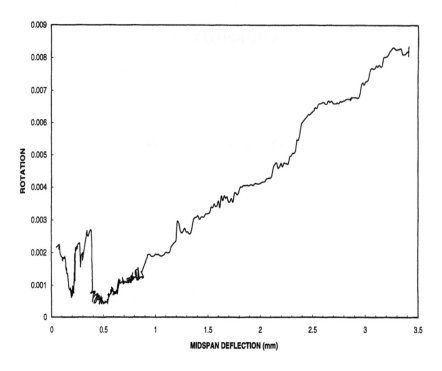

A_012_06#3

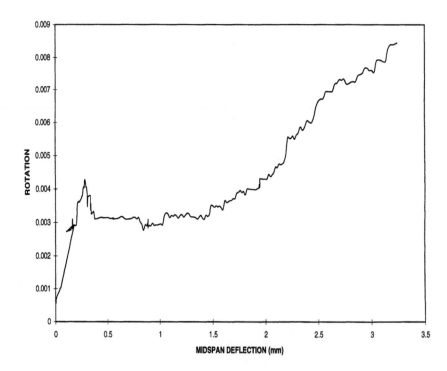

A_025_06#1

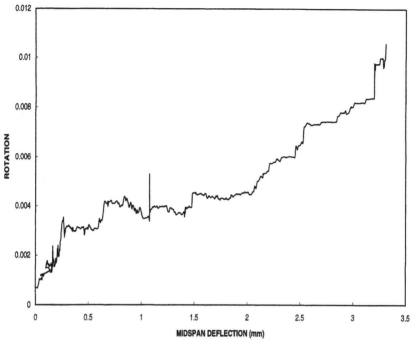

A_025_06#3

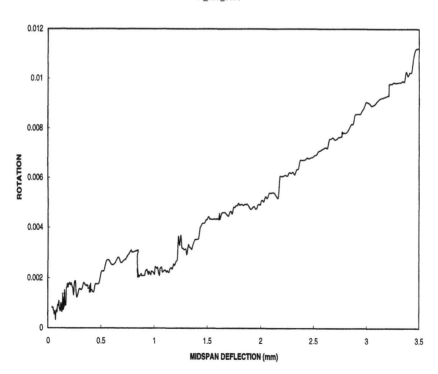

D. Lange-Kornbak and B. L. Karihaloo

A_088_06#1

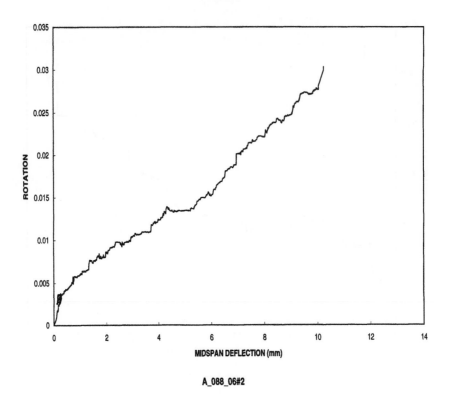

A_088_06#2

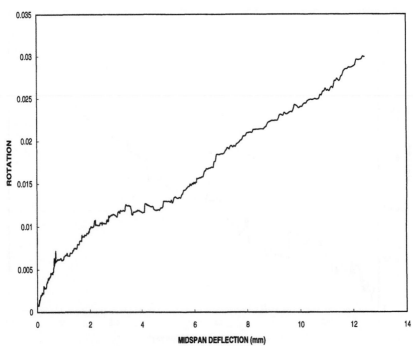

A_088_06#3

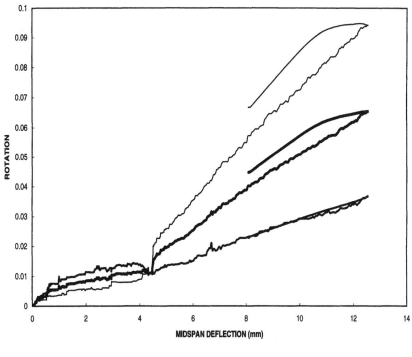

B_012_12#1

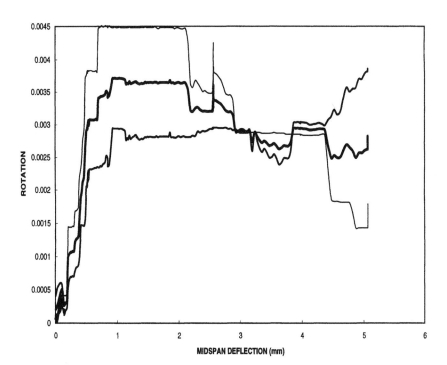

B_012_12#2

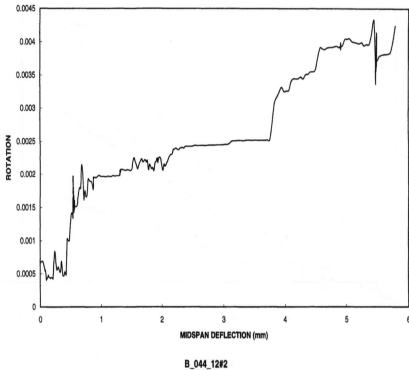

B_044_12#2

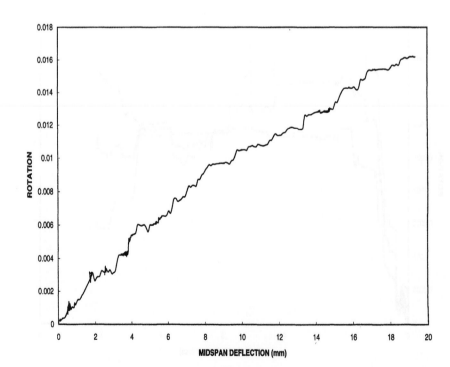

B_200_12#1

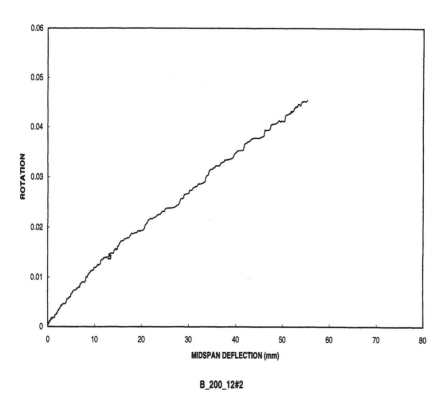

B_200_12#2

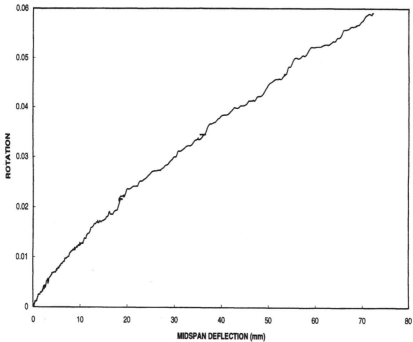

APPENDIX C

FAILURE MODES

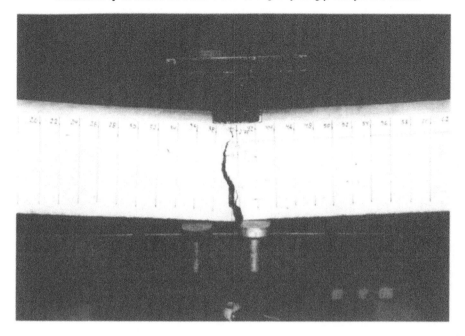

A_012_06 #1

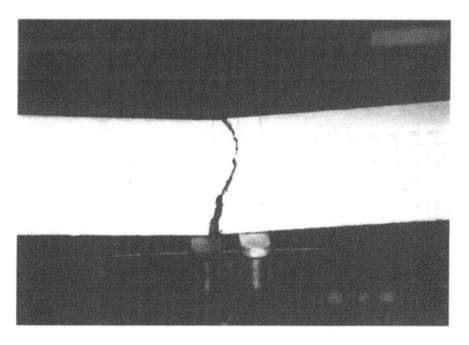

A_012_06 #2

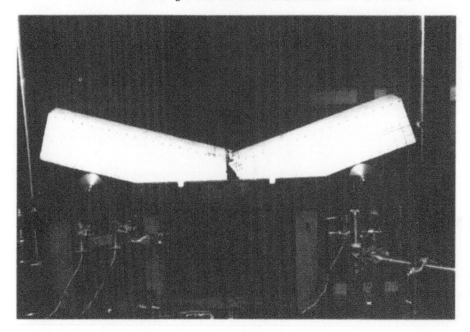

A_012_06 #3

A_025_06 #1

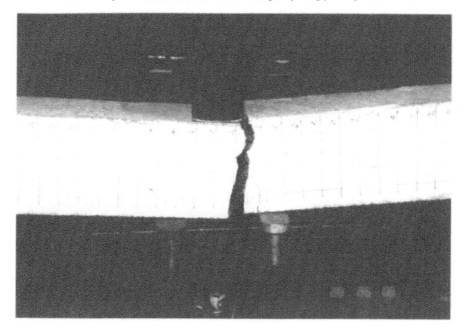

A_025_06 #2

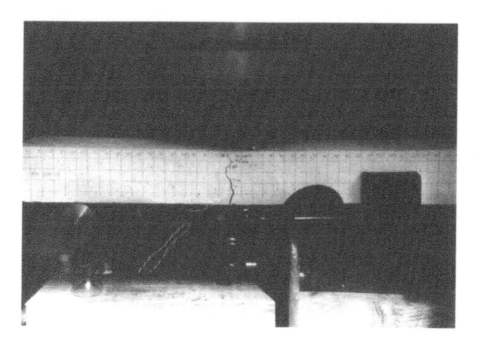

A_025_06 #3

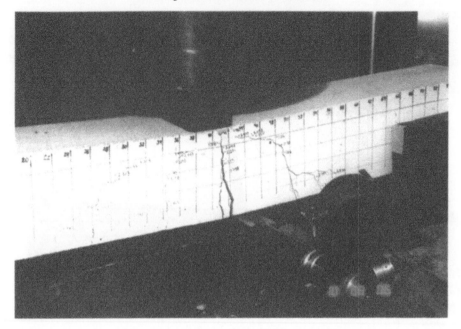

A_088_06 #1

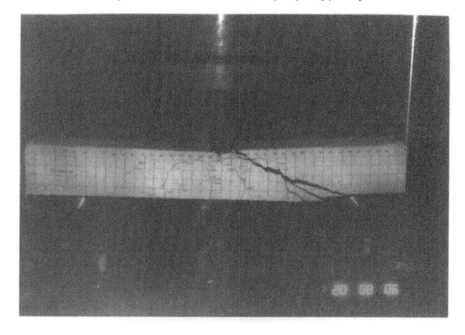

A_088_06 #2

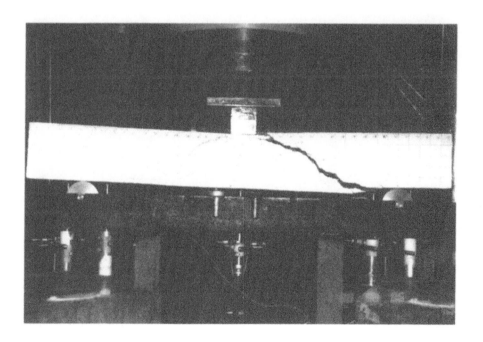

A_088_06 #3

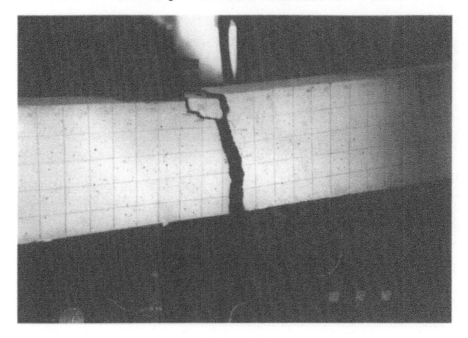

B_012_12 #1

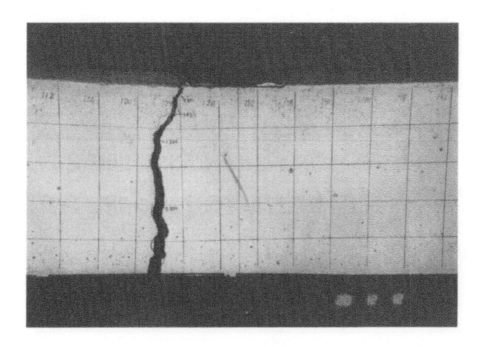

B_012_12 #2

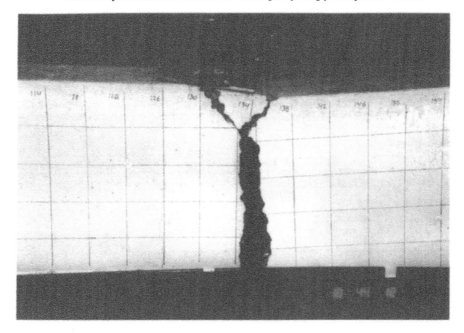

B_044_12 #1

B_044_12 #2

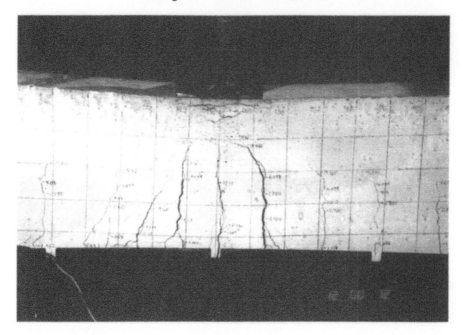

B_200_12 #1

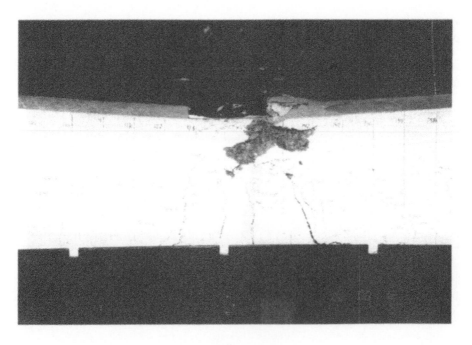

B_200_12 #2

SIZE EFFECT AND BOND-SLIP DEPENDENCE OF LIGHTLY REINFORCED CONCRETE BEAMS

G. RUIZ, M. ELICES and J. PLANAS

Departamento de Ciencia de Materiales, Universidad Politécnica de Madrid,
E. T. S. de Ingenieros de Caminos, Ciudad Universitaria,
28040 Madrid, Spain

ABSTRACT

This contribution presents the essentials of a research program, both theoretical and experimental, designed to improve the understanding of the mechanical behavior of lightly reinforced beams, particularly their transition from brittle to ductile behavior. A simplified model — called the effective slip-length model— describes the concrete fracture as a cohesive crack and incorporates the effect of reinforcement bond-slip. A numerical algorithm is described which is computationally efficient, and so specially suited for parametric analyses of the problem. The experimental research is conducted on three-point bend, lightly reinforced microconcrete beams. Although the beams were of reduced size, the properties of the microconcrete were selected so that the behavior observed is representative of beams of ordinary size made of ordinary concrete. The experiments study the effect of steel ratio, beam depth and bond strength, and include the determination by independent tests of all the parameters of the model. The numerical predictions of the experimental results by the effective slip-length model are reasonably good. The model is able to capture minute experimental details, such as a secondary peak in the load for relatively large steel covers, and describes well the transitional behavior of lightly reinforced beams from brittleness to ductility. Based on numerical analyses, a closed-form expression is given for minimum reinforcement in bending which is compared to recommendations from building codes and formulas from other authors. The comparison shows that the recommendations in the codes could be improved to get safer or cheaper minimum reinforcement to avoid brittle behavior. It also shows that the influence of bond strength should be taken into account, and that generally speaking larger bond strength requires larger reinforcement, although the quantitative effect of the bond strength depends on the details of the beam, particularly on the reinforcement cover, and on the concrete and steel grades.

KEYWORDS

Size effect, bond-slip, minimum reinforcement, cohesive crack model, fracture energy.

INTRODUCTION

For most ordinary beam design the bending failure of reinforced concrete beams is ductile, but in practice it may occur that beam dimensions are dictated by criteria other than the strictly mechanical, and then beams may require very little reinforcement for a strict application of ultimate strength design (yield of steel and no tension for concrete). When the tensile strength of concrete is taken into account, the cracking strength of the beam may be larger than the ultimate strength, which can lead to brittle behavior,

67

i.e., once the load reaches the cracking strength, collapse is instantaneous because this load is larger than the ultimate (plastic) strength. Thus, the behavior of lightly reinforced beams shows a transition from a brittle behavior, at very low reinforcement ratios, to ductile behavior (ultimate strength larger than the cracking strength) at reinforcement ratios above a certain threshold, called the minimum reinforcement.

Since the transition from brittle to ductile is determined by the cracking process of concrete, its analysis is based on fracture mechanics concepts, the results depending on the structure size (beam depth in this case). However, in the building codes, the minimum reinforcement is defined from concepts of the theory of plasticity, which leads to minimum reinforcement ratios independent of the beam dimensions, with no consideration of size effect. In addition, in determining minimum reinforcement, the codes only consider the concrete compressive strength, and focus on the way it affects the beam strength, not on its brittleness or ductility. ESIS TC9 was interested in this subject and promoted experimental research [1-4], and several theoretical analyses based either on linear elastic fracture mechanics [3, 5-7] or on cohesive crack concepts [4, 8-13].

Among the foregoing references, only the works of Hededal and Kroon [4], of Ruiz and Planas [12] and of Ruiz [13] consider the possibility of bond-slip between reinforcing bars and concrete, while these works show that bond-slip does substantially influence the response of the beams. On the other hand, most of the available experimental programs do not provide independent measurement of all the properties of the materials relevant to the analysis, particularly the bond-slip properties. Thus, the need was felt for an experimental research with an independent determination of all the properties of concrete, steel and steel-concrete interface, as well as tests of lightly reinforced beams of various sizes, with various reinforcement ratios and bond strength. An outline of the most important results of this research was given by Planas, Ruiz and Elices [14]. This chapter presents a broader description of the experimental results and discusses them from an approach based on the cohesive crack model that takes account of the bond-slip properties.

The chapter is structured as follows: an outline of the fundamentals of the model is given in the second Section, wich follows this introduction. The third Section gives a brief description of the numerical procedure and an analysis of the sensitivity of the numerical results to various parameters, such as the beam depth, the reinforcement ratio and the steel-to-concrete bond properties. The fourth Section describes the materials, the specimens and the experimental program, and compares the tests results with the predictions of the model. The fifth Section discusses the sensitivity of the experimental results to the parameters influencing the theoretical model, and compares a minimum reinforcement formula, calculated with this model, with other available formulas, both from building codes and from fracture analyses by other authors. The final Section summarizes the main points of the research.

A MODEL FOR LIGHTLY REINFORCED CONCRETE BEAMS

This section explains the fundamentals of a theoretical model devised to analyze the behavior of a lightly reinforced concrete beam with a single layer of reinforcement, when subjected to three-point bending. The model assumes various hypotheses concerning three main aspects of the problem: (1) failure mechanism, (2) concrete and steel behavior, and (3) the bond between steel and concrete.

Failure Mechanism

The model assumes the following basic hypotheses:

H.1 *In a lightly reinforced beam loaded in three-point bending, only one crack progresses, which is located at the central cross section* (Fig. 1a). This in fact *defines* light reinforcement, and therefore if multiple cracking occurs the beam is no longer lightly reinforced.

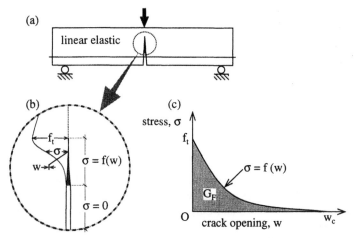

Fig. 1. (a) Reinforced beams subjected to three-point bending; (b) detail of the cohesive crack; (c) softening function.

H.2 *The crack is cohesive: stresses are transferred between the crack faces.*

A cohesive crack (Fig. 1b) initiates where the maximum principal stress reaches the tensile strength of the material, f_t. The crack is perpendicular to the direction of the maximum principal stress and opens transferring stress from one face to another. For monotonic mode I opening, the stress transferred is normal to the crack faces and is a single function of the crack opening :

$$\sigma = f(w), \quad \text{and} \quad \begin{cases} f(0) = f_t \\ f(w) \geq 0 \end{cases} \tag{1}$$

The function $f(w)$ is known as the *softening function* and is considered characteristic of the material. The values of $f(w)$ become zero when the crack opening exceeds a critical value w_c (Fig. 1c).

The *specific fracture energy*, G_F, is the work needed to separate the two faces of a unit surface area of crack, and is represented by the area under the softening curve (Fig. 1c). Since the softening function is a material property, G_F is also a constant material property. Another important parameter is the *characteristic length*, l_{ch}, which characterizes the intrinsic brittleness of the concrete [15] and is defined as:

$$l_{ch} = \frac{E\, G_F}{f_t^2} \tag{2}$$

The cohesive crack is implemented in this model using the smeared-tip method developed by Planas and Elices [16, 17], and later adopted, with some modifications, by Bazant, and Bazant and Beisel [18, 19]. A version including internal stresses was later developed by Planas and Elices [20, 21] which is explained in detail below.

Material Behavior

The second group of hypotheses refers to the behavior of concrete and steel:

H.3 *The concrete behavior outside the fracture zone is linear elastic.* This assumption implies only a second order error, because in a lightly reinforced beam the inelastic strains before cracking are small. This assumption is not conceptually essential, and a more complex bulk behavior could be

assumed. However, bulk linearity allows the use of numerical methods which are efficient and could not be used if nonlinear behavior were assumed.

H.4 *The steel is considered as elastic-perfectly plastic.* This hypothesis is introduced for simplicity, and also to facilitate the observation of the behavior of the ligthly reinforced beam when the steel is in the elastic or in the plastic regime. The model could be modified to implement any stress-strain law for the steel.

Bond between Steel and Concrete

The last group of hypotheses refers to steel-to-concrete interaction and how it is used to analyze the reinforced beam behavior. The simplest model, widely used in reinforced concrete analysis, assumes perfect bond; it was adopted by Bosco and Carpinteri in their analysis of lightly reinforced concrete beams based on LEFM [5, 6]. The steel action is then equivalent to two closing forces acting on the crack faces at the point where the reinforcement crosses the crack. While the steel remains elastic, the force is determined by the condition that the crack opening is zero at the point where the reinforcement crosses the crack. When the steel yields, the closing forces become constant, equal to the yield force of the reinforcing bars (which means that perfect plasticity is assumed for the reinforcement). Hededal and Kroon [4] consider bond-slip assuming that the shear stress-slip law is of the rigid-plastic type, but replace the shearing forces by their resultant applied on the crack lips.

These two approximations give a behavior stiffer than the real one, because the closing forces act on the crack lips, while the real stresses are distributed along the contact with the reinforcement. Moreover, if the bond-slip is prevented, the compliance before steel yielding is greatly reduced, and if reinforcement forces are applied directly on the crack lips, the beam response depends on the size of the area on which the forces are distributed. Massabò [7] used a version of Bosco and Carpinteri's model in which the forces were distributed over an area proportional to the diameter of the steel bars. Hededal and Kroon apply the forces concentrated at the finite element mesh nodes at the level of the reinforcement. This implies that the solution depends on the element size [13].

To solve these difficulties, the authors replaced the shear stresses acting along the contact with the reinforcement by their resultant force applied at a point *inside* the concrete. The entire bond-slip behavior of the model rests on the following hypotheses:

H.5 *The shear stresses τ transferred between reinforcement and matrix is a function of the relative slip, s, between the two materials.* As a simple and reasonable approximation, this function is taken as rigid-perfectly plastic.

The model assumes that the steel-concrete interaction can be modeled by means of a bond shear stress-slip law [22, 23]. Further experimental evidence suggests that the the axial stress law in steel bars is distributed according to a triangular law, which implies that a constant bond shear stress is acting on the contact surface [24, 25], so a rigid-perfectly plastic shear stress-slip law can be assumed as a simple and reasonable approximation:

$$\begin{cases} \tau \leq \tau_c & \text{if } s = 0 \\ \tau = \tau_c & \text{if } s \neq 0 \end{cases} \tag{3}$$

where τ is the shear stress at the steel-concrete interface, τ_c is the bond shear strength, and s is the relative slip between materials. This τ-s law was proposed by Bazant and Cedolin [22] and the ACI Committee 446 [26] and leads to simple analytical solutions. It was used to analyze lightly reinforced beams by Hededal and Kroon [4]. It is not, however, essential to the model, and more complex stress-slip laws could be applied.

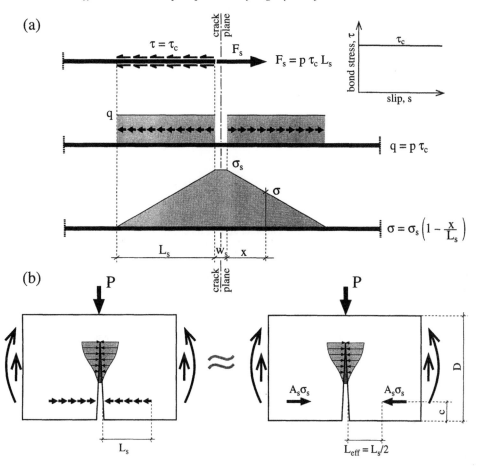

Fig. 2. (a) Stresses on the reinforcement bars, and (b) mechanical actions due to the reinforcement on the concrete beam, that are replaced by a pair of closing forces.

H.6   *The actual bond shear stress distribution can be replaced by a pair of concentrated forces acting* inside *the concrete*. The location of these forces is determined by enforcing mechanical equivalence with the actual (continuous) stress distribution, in a way similar to that used by Bazant and Cedolin [27] in another context.

    To obtain an explicit formulation of the model, consider the reinforcement isolated from concrete as depicted in Fig. 2a. The protrusion of the steel bars coincides with the crack opening at the point where the crack crosses the reinforcement layer, w_s, and at that point the steel is loaded by $F_s = A_s \sigma_s$. We assume that the steel cross-section A_s is a small fraction of the concrete cross-section, so the concrete can be considered as rigid in this context. Therefore, the steel is essentially unstrained except over the slip zones, of length L_s (see Fig. 2a).

    According to H.5, the shear stress is constant over the slip length, and the equilibrium of the horizontal forces acting on an arbitrary portion of the reinforcement requires that $A_s \sigma(x) = \tau_c\, p\,(L_s - x)$, where p is the perimeter of the cross section of the reinforcement. Global equilibrium further requires that $F_s = \sigma_s A_s = \tau_c\, p\, L_s$ from which it follows that the slip length is given by

$$L_s = \frac{A_s \, \sigma_s}{\tau_c \, p} \tag{4}$$

The relationship between the force and w_s is obtained by integrating the strain along the slip length:

$$w_s = 2 \int_0^{L_s} \varepsilon[\sigma(x)] \, dx \tag{5}$$

where $\varepsilon(\sigma)$ is the strain (uniaxial) on the steel under the stress σ. For an elastic-perfectly plastic behavior of the steel, it follows from (5) and (4) that the relationship between the force F_s and the crack opening at the level of the reinforcement w_s is:

$$F_s = A_s \, \sigma_s = \begin{cases} A_s \left(E_s \tau_c \dfrac{p}{A_s} \right)^{1/2} \sqrt{w_s} & \text{if } w_s < w_y = \dfrac{A_s f_y^2}{E_s \tau_c p} \\[2mm] A_s f_y & \text{if } w_s > w_y \end{cases} \tag{6}$$

where f_y is the yield stress of the reinforcement steel and E_s its elastic modulus.

The foregoing model is implemented in the lightly reinforced beam as depicted in Fig. 2b where the shear stresses extend over the slip length L_s at both sides of the crack. As a further simplification, the shear forces can be replaced by two forces *concentrated* at points located at a certain *effective slip length* L_{eff} at both sides of the crack (hypothesis H.6). This is equivalent to substituting the actual reinforcement by an *unbonded* bar of length $2 L_{eff}$ anchored in the concrete at its ends. The effective slip length is calculated to keep the relation between F_s and w_s identical to that of the original model. For a linear-elastic response of steel, this turns out to be equivalent to replacing the shear stress distribution by its resultant applied at the center of gravity of the distribution, which for the uniform distribution is located at the mid point of the slip length, i. e. $L_{eff} = L_s/2$. From (6) it follows that the effective slip length is given, in terms of the crack opening w_s by

$$L_{eff} = \begin{cases} \sqrt{\dfrac{A_s E_s}{4 \, \tau_c \, p}} \, w_s & \text{for} \quad w_s < w_y = \dfrac{A_s f_y^2}{E_s \tau_c p} \\[3mm] \dfrac{A_s \, f_y}{2 \, \tau_c \, p} & \text{for} \quad w_s > w_y \end{cases} \tag{7}$$

The foregoing reduction of the action of the reinforcement on the concrete to a pair of closing forces, one at each side of the crack, has the virtue of being amenable to analytical treatment, which makes it possible to implement them in a finite element based program using the smeared-tip method, which was originally developed to study the fracture behavior of plain concrete beams with internal stresses due to shrinkage or creep. The next section describes the fundamentals of this method and the way the pair of closing forces are taken into account.

NUMERICAL RESULTS

The propagation of a crack through the central section of a lightly reinforced beam loaded at three points can be solved numerically by using a simple algorithm called the method of the smeared-tip with internal stresses. This method has been successfully used to study the cracking process of plain concrete beams with internal stresses generated by shrinkage or temperature gradient across the specimen [20, 21]. In the present case the internal stresses are those generated by the reinforcement in the middle cross section of an uncracked beam. This section explains the fundamentals of the numerical procedure (Section 3.1) and gives a brief analysis of the effect of material and geometrical parameters on the numerical predictions of the model described in the previous section (Section 3.2).

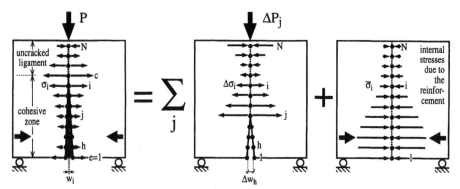

Fig. 3. Superposition of elastic cases used in the smeared-tip method.

Calculation Method: Smeared -Tip with Internal Stresses

In the smeared-tip method, the solution is sought as the superposition of elastic cases as depicted in Fig. 3. The stresses and displacements for the actual nonlinear state (Fig. 3, left) are written as the sum of N elastic cases corresponding to N different crack lengths (Fig. 3, middle), plus a further elastic case corresponding to the uncracked body subjected to the internal stresses. The load, the displacement under the load-point, the nodal stresses and the nodal crack openings can then be written as:

$$P = \sum_{j=1}^{N} \Delta P_j \tag{8}$$

$$\delta = \sum_{j=1}^{N} C_j \Delta P_j + \overline{\delta} \tag{9}$$

$$\sigma_i = \sum_{j=1}^{N} R_{ij} \Delta P_j + \overline{\sigma}_i \qquad i = 1, 2, ..., N \tag{10}$$

$$w_i = \sum_{j=1}^{N} D_{ij} \Delta P_j \qquad i = 1, 2, ..., N \tag{11}$$

where C_j is the displacement under the load-point generated by an external unit force for a crack with its tip located at node j, $\overline{\delta}$ is the load-point displacement generated by the internal stress field, R_{ij} are the elements of the reaction matrix $\mathbf{R}$, and D_{ij} the elements of the displacement matrix $\mathbf{D}$, defined as follows.

The element R_{ij} of the matrix $\mathbf{R}$ is the reaction force at node i when a unit external force is applied with a crack reaching node j. Consequently, all elements where $i < j$ are zero:

$$R_{ij} = 0 \quad \text{if} \quad i < j \tag{12}$$

This means that $\mathbf{R}$ is a lower triangular matrix, i. e.,

$$\mathbf{R} = \begin{bmatrix} R_{11} & 0 & 0 & ... & 0 & 0 \\ R_{21} & R_{22} & 0 & ... & 0 & 0 \\ ... & ... & ... & ... & 0 & 0 \\ R_{N-1\,1} & R_{N-1\,2} & R_{N-1\,3} & ... & R_{N-1\,N-1} & 0 \\ R_{N1} & R_{N2} & R_{N3} & ... & R_{N\,N-1} & R_{NN} \end{bmatrix} \tag{13}$$

The element D_{ij} in matrix $\mathbf{D}$ is the crack opening at node i when a unit force is applied with a crack reaching node j. Consequently, all elements where $i \geq j$ are zero:

$$D_{ij} = 0 \quad \text{if } i \geq j \tag{14}$$

So **D** is an upper triangular matrix with zero diagonal elements, i. e.

$$\mathbf{D} = \begin{bmatrix} 0 & D_{12} & D_{13} & \cdots & D_{1\,N\text{-}1} & D_{1N} \\ 0 & 0 & D_{23} & \cdots & D_{2\,N\text{-}1} & D_{2N} \\ 0 & 0 & 0 & \cdots & \cdots & \cdots \\ 0 & 0 & 0 & \cdots & 0 & D_{N\text{-}1\,N} \\ 0 & 0 & 0 & \cdots & 0 & 0 \end{bmatrix} \tag{15}$$

R, **D** and C_j can be calculated once and for all for a given geometry, and stored in a full matrix **M** of dimensions $N \times (N+1)$ as

$$\mathbf{M} = \begin{bmatrix} R_{11} & D_{12} & D_{13} & \cdots & D_{1\,N\text{-}1} & D_{1N} & C_1 \\ R_{21} & R_{22} & D_{23} & \cdots & D_{2\,N\text{-}1} & D_{2N} & C_2 \\ \cdots & \cdots & \cdots & \cdots & \cdots & \cdots & \cdots \\ R_{N\text{-}1\,1} & R_{N\text{-}1\,2} & R_{N\text{-}1\,3} & \cdots & R_{N\text{-}1\,N\text{-}1} & D_{N\text{-}1\,N} & C_{N\text{-}1} \\ R_{N1} & R_{N2} & R_{N3} & \cdots & R_{N\,N\text{-}1} & R_{NN} & C_N \end{bmatrix} \tag{16}$$

To write the equations governing the actual nonlinear problem, we consider that the cohesive zone has extended up to node c (Fig. 3, left), and use (10) and (11) to write that the stresses and crack openings over the cohesive zone satisfy the softening equation $\sigma_i = f(w_i)$ —Eq. (1)—, and that on the uncracked ligament $w_i = 0$. The results of the equations are, respectively,

$$\sum_{j=1}^{N} R_{ij}\,\Delta P_j + \bar{\sigma}_i = f\left(\sum_{j=1}^{N} D_{ij}\,\Delta P_j\right) \qquad i = 1, 2, \ldots, c \tag{17}$$

$$\sum_{j=1}^{N} D_{ij}\,\Delta P_j = 0 \qquad i = c, c+1, \ldots, N \tag{18}$$

where $\bar{\sigma}_i$ is the internal stress caused by the reinforcement at node i; $f(w)$ is the softening function of the cohesive material (1); and the cohesive crack extends up to the node c.

To solve the system we start with equation (18), whose solution is trivial because the system is triangular:

$$\Delta P_j = 0 \qquad \text{for} \qquad j = c+1, c+2, \ldots N \tag{19}$$

Inserting this result in (17) and rearranging, we get

$$\sum_{j=1}^{c} R_{ij}\,\Delta P_j = f\left(\sum_{j=1}^{c} D_{ij}\,\Delta P_j\right) - \bar{\sigma}_i \qquad i = 1, 2, \ldots, c \tag{20}$$

This is a system of c equations for the c unknowns ΔP_j ($j = 1,\ldots, c$), which can be solved iteratively as follows: starting from an estimate for ΔP_j, the right-hand member is evaluated, and the system is solved for a better estimate. Since **R** is triangular, the iteration is very fast. At each iteration the internal stresses $\bar{\sigma}_i$ introduced by the reinforcement must be evaluated. Let us remember that these $\bar{\sigma}_i$ are the stresses generated at both sides of the crack, at the central cross section of an uncracked elastic beam (Fig. 3, right) by the closing forces F_s given by (6), placed at a distance L_{eff} given by (7). The detailed procedure followed to get the resulting stress distibution is described in Appendix A. Here it is enough to point out that the stress distribution can be written as

$$\bar{\sigma} = \rho \frac{F_s}{A_s} F'(D, y, c, L_{eff}) \tag{21}$$

where ρ ($\rho = A_s/A_c$) is the steel ratio, D is the beam depth, y is the distance from the actual point in the central cross section to the bottom of the beam, and c is the depth of the reinforcement concrete cover. $F'(\cdot)$ is a dimensionless function given in Appendix A.

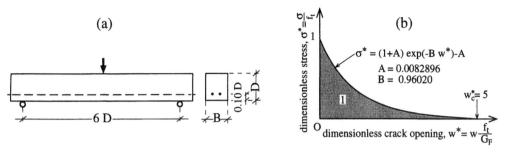

Fig. 4. (a) Beam dimensions, and (b) softening function used in the calculations to study the influence of various parameters.

Influence of Material and Geometrical Parameters

To disclose the effect of the various material and geometrical parameters involved in the problem, various computational runs were performed to see their relative influence. To draw the resulting curves in a nondimensional form for direct comparison, the following dimensionless load and deflection were defined (identified by a star):

$$P^* = \frac{\sigma_N}{f_t} = \frac{3}{2}\frac{P}{BD^2 f_t}\frac{s}{} \quad , \quad \delta^* = \delta\frac{f_t}{G_F} \tag{22}$$

The relevant parameters are the relative stiffness between steel and concrete, the beam depth (size effect), the reinforcement ratio together with its yield strength, and the bond properties. These parameters can be expressed in nondimensional form as follows:

$$n = \frac{E_s}{E_c} \quad , \quad \beta_H = \frac{D}{l_{ch}} \quad , \quad \rho = \frac{A_s}{A_c} \quad , \quad f_y^* = \frac{f_y}{f_t} \quad , \text{ and } \quad \eta = \left(n\,\frac{\tau_c}{f_t}\frac{p\,l_{ch}}{A_s}\right)^{1/2} \tag{23}$$

where 'n' is the relative stiffness between steel and concrete; β_H is known as the Hillerborg's brittleness number and represents the beam depth in a nondimensional form (β_H is one essential parameter in the description of fracture processes in plain concrete [28, 29]); ρ is the reinforcement ratio; f_y^* is the nondimensional steel yielding strength and η is a nondimensional bond parameter which appears when the closing force F_s in Eq. (6) is reduced to nondimensional form. It should be noted that η depends mainly on the bond strength τ_c, on the rebars perimeter p and on the relative stiffness 'n', while the remaining parameters (f_t, l_{ch} and A_s) are introduced to make it nondimensional.

We next present some numerical results to show how these parameters influence the behavior of a reinforced beam with the dimensions shown in Fig. 4a. All the calculations were performed using the quasi-exponential softening function sketched in Fig. 4b [30].

Influence of the Relative Stiffness 'n'. Figure 5 shows dimensionless load-displacement curves in which only the relative stiffness 'n' is varied while keeping β_H, ρ, η and f_y^* constant. The results show that the initial elastic response and the plastic branch are essentially unaffected by varying 'n' in the range 4 to 13. The effect is noticeable in the intermediate range, in which the concrete cracks while the steel remains elastic, but it is very small for practical purposes, since the maximum relative effect is of the order of 2% (at the relative minimum) while the effect on the peak load is only 0.4%.

So it can be concluded that the influence of the steel-to-concrete relative stiffness 'n' is very small by itself, although it can influence the numerical response of the model through the adherence parameter η.

Influence of the Size. Figure 6 shows nondimensional P-δ curves obtained for four geometrically proportional beams (only the size varies, the rest of the parameters remain constant; in Fig. 6, δ is also divided by β_H to get identical initial slopes). As β_H increases, the nondimensional strength decreases,

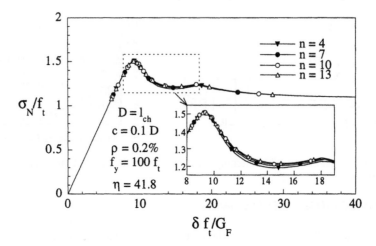

Fig. 5. Variation of a nondimensional P-δ curve when the moduli ratio 'n' varies within the usual range between concrete and steel $(4 \leq n \leq 13)$.

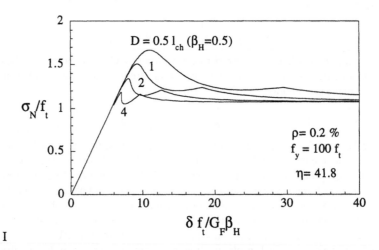

I

Fig. 6. Nondimensional P-δ curves for geometrically similar lightly reinforced beams.

and the beam response is weaker for bigger beams, which means that there is size effect in the response of this model.

Influence of the Reinforcement Ratio. Figure 7 shows nondimensional P-δ curves obtained for a beam with various reinforcement ratios. The reinforcement ratio increases the ultimate strength of the beams proportionally to the dimensionless group ρf_y^*, while the response of the intermediate zone of the P-δ curve is dependent on the intensity of the closing forces generated by the reinforcement, which depends directly on the product of the reinforcement ratio by the nondimensional adherence, $\rho \eta$.

Influence of Bond Properties. To study the influence of the bond strength on the model response, several calculations were done on the same beam, varying only the parameter η. Fig. 8a shows the results for a beam reinforced with linear elastic reinforcement without plastic branch $(f_y = \infty)$: for stronger bonds the peak load is larger, and the response after the peak is stronger and stiffer.

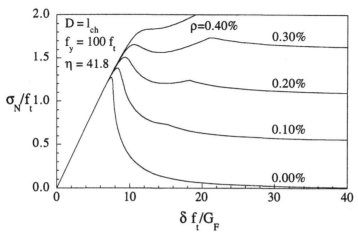

Fig. 7. Nondimensional P-δ curves for several reinforcement ratios.

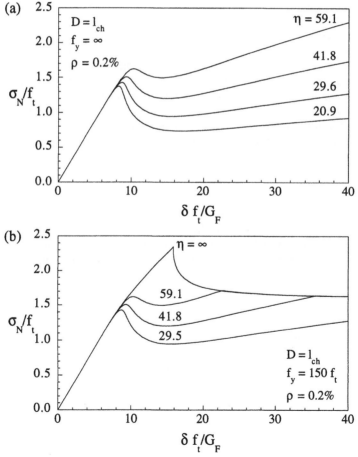

Fig. 8. Nondimensional P-δ curves for reinforced beams varying the steel-to-concrete bond and the steel yielding strength: (a) $f_y = \infty$; and (b) $f_y = 150\, f_t$.

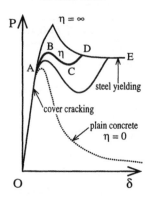

Fig. 9. Scheme of the variation in the P-δ curve as a function of the bond parameter.

Fig. 8b shows nondimensional P-δ curves for beams of the same characteristics (in particular, the same depth and reinforcement ratio) but for which the reinforcement yields at $f_y = 150 f_t$. These curves have two well differentiated branches. In the first branch the reinforcement remains elastic and the curves are identical to those of Fig. 8a. The second branch, after reinforcement yielding, is almost independent of the bond strength, and is mainly controlled by the reinforcement ratio and yield strength.

Fig. 9 sketches the general trends of the beam response. Along the first branch (OABD) the reinforcement remains elastic and the response is sensitive to the reinforcement ratio ρ and to the adherence η. For a given ρ, this part of the curve lies between two bounding curves: the upper bound corresponds to η = ∞ (perfect bond between reinforcement and matrix) and the lower bound to η = 0 (free unbonded reinforcement; this limiting curve is identical to that for a plain concrete beam). This phase ends when the reinforcement starts to yield, and then the curve follows the second branch (DE in Fig. 9) which is essentially controlled by the mechanical resistance of the reinforcement (ρ f_y^*). Its decreasing trend is due to the lower contribution from concrete to the overall strength as failure advances.

EXPERIMENTAL RESULTS

The experiments were designed to investigate the dependence of the fracture of lightly reinforced concrete beams on the specimen size and on the bond of steel to concrete, apart from such well known parameters as the steel strength and the steel ratio. In addition, the program had to provide all the data needed for numerical simulations. These properties have to be measured from independent tests.

Testing Program

Two basic conditions guided the design of the lightly reinforced beams. First, a minimum of three widely different sizes were to be tested to analyze size effect; the reinforcement ratio and the bond properties were also to be varied to study their influence on the beam behavior. Second, the dimensions and weight of the laboratory beams had to allow easy handling, and their behavior be representative of the behavior of beams of ordinary size made of normal concrete.

Geometrically similar beams of three sizes, three steel ratios and two different bond properties were designed. Figure 10a sketches the selected beam geometry and loading conditions, and Fig. 10b the basic properties of each type of beam (depth, D, steel ratio, ρ, smooth or ribbed bar, S or R) and the number of specimens of each type. A relatively brittle microconcrete —l_{ch} = 130 mm— was selected. In this way, the relatively small laboratory beams display the same fracture behavior as ordinary concrete beams two or three times bigger, the value of l_{ch} for ordinary concrete being about 300 mm.

Standard characterization and control tests were performed to determine the compressive strength, tensile strength, elastic modulus and fracture energy of the concrete. Steel properties and steel-to-concrete bond characteristics were also measured. The yield strength and the elastic modulus of the reinforcing steel were determined from tensile tests. Bond strength of the steel-concrete interface was determined from pull-out tests in which both the pulling force and the steel slip were measured.

Materials. A single micro-concrete mix was used throughout the experimentation, composed of a siliceous aggregate of 5 mm maximum size and rapid hardening Portland cement (ASTM type III). The proportions by weight were $3:0.5:1$ (aggregate:water:cement), with a cement content of 500 kg/m³. The granulometric curve of the aggregate follows ASTM C33, and was achieved by appropriate mixing of 4 aggregate fractions. All the cement used was taken from the same cement container, and dry-stored until use. All the specimens were made from 6 batches of 80 liters each. The first batch was used to make the characterization specimens, while the reinforced beams and various control specimens were made from the remaining 5 batches.

The control specimens were tested to check that the properties from the various batches were close enough. The Abrams cone slump was measured immediately before casting. All the specimens were cast in steel molds, vibrated during 10 seconds fastened to a vibrating table, wrap-cured for 24 hours, demolded, and stored in a moist chamber at $20 \pm 2\,°C$ for three weeks and then under water until testing time (8 weeks). Table 1 shows the characteristic parameters of the micro-concrete used throughout the experimentation, determined in the various characterization and control tests.

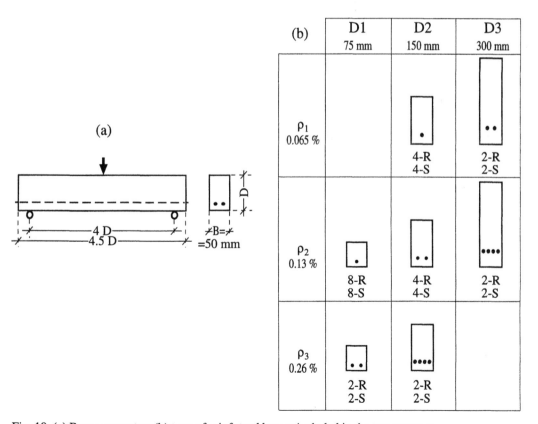

Fig. 10. (a) Beam geometry; (b) type of reinforced beams included in the test program.

Table 1. Micro-concrete characteristics (8 weeks).

	slump mm	$f_c^{(a)}$ MPa	$f_{ts}^{(b)}$ MPa	$E_c^{(a)}$ GPa	$E_{C\,SEN}^{(c)}$ GPa	G_F N/m	l_{ch} mm
mean	120	39.5	3.8	30.5	30.3	62.5	130
std. dev.	12	1.6	0.34	2.0	3.2	4.7	—

(a) Cylindrical specimens, compression. (b) Cylinder splitting (Brazilian).
(c) From initial CMOD compliance of notched beams.

Table 2. Properties of the reinforcement steel bars

Wire type	E_s (GPa)	σ_u (MPa)	$\sigma_{0.2}$ (MPa)	ε_r (%)
smooth	200 ± 4	608 ± 5	568 ± 6	3.5
ribbed	162 ± 8	587 ± 2	538 ± 2	2.3

Table 3. Reinforced beam characteristics

Denomination [a]	D1-R2X	D1-R3X	D2-R1X	D2-R2X	D2-R3X	D3-R1X	D3-R2X
D (mm)	75	75	150	150	150	300	300
number of wires	1	2	1	2	4	2	4
steel ratio, ρ(%)	0.130	0.260	0.065	0.130	0.260	0.065	0.130
No. specimens [b]	8+8	2+2	4+4	4+4	2+2	2+2	2+2

(a) 'X' is a place holder for 'S' = smooth wire, or 'R' = ribbed wire.
(b) Half with smooth wire + half with ribbed wire.

Commercial wires with a nominal diameter of 2.5 mm were used to achieve the desired steel ratios. All the wires were originally smooth, so to get stronger bond, v-shaped ribs were stamped on half of them at a 2 mm spacing. The reduction in diameter at the deepest point of the ribs was 7.5%, with 10% protrusions on either side of the rib. Table 2 shows the elastic modulus E_s, the ultimate strength σ_u, the 0.2% offset yield strength $\sigma_{0.2}$, and the ultimate strain, ε_r, for both smooth and ribbed wires. The nominal value of the diameter was used to determine the parameters shown in the table.

Characterization and Control Specimens. Cylindrical specimens 150 mm in length and 75 mm in diameter were cast in metallic molds, 12 for compression tests and 12 for splitting tests, 4 specimens from each batch.

Notched plain concrete beams were used to characterize concrete fracture properties. All the beams were 50 mm in thickness and their length was 4.5 times their depth; 24 beams 75 mm in depth were cast for characterization and control, 4 specimens from each batch. 4 beams 150 mm in depth, and 2 beams 300 mm in depth were cast from the first batch to check size effect. Notches were sawn before testing at the central cross-section to a depth of half the total beam depth ($a_0/D = 0.5$).

Pull-out specimens were cast, consisting of prisms $75 \times 75 \times 150$ mm with a wire embedded along their longitudinal axis, 10 for each kind of wire, smooth or ribbed. These specimens were prepared from the batches from which the reinforced beams were made.

Reinforced Micro-Concrete Beams. Table 3 and Fig. 10 show the dimensions and reinforcement ratio of the reinforced concrete beams. The specimens were cast in metallic molds, with the reinforcing wires protruding through holes at the ends of the mold walls. The micro concrete was compacted on a vibrating table. During casting and vibration the wires were tensioned by nuts to hold them tight and in place. Just after the end of the manipulation, the tension of the wires was released. After demolding, the wires were left protruding from the end of the beam, as sketched in Fig. 10a. No hooks or anchors were used.

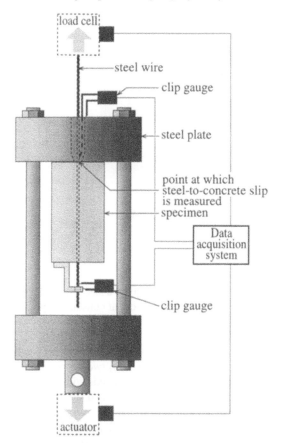

Fig. 11. Experimental set-up for the pull-out tests.

Experimental Procedures. Two set of tests were carried out: (1) characterization and control tests, and (2) reinforced beam tests, which are described next.

Characterization and control tests. Compression tests were carried out on 12 cylindrical specimens —two from each batch— according to ASTM C-39 and C-469 (except for a reduction in size). The strain was measured over a 50 mm gage length by means of two clip gauges placed symmetrically. The tests were run under displacement control, at a rate of 0.3 mm/min.

Brazilian tests were also carried out on 12 cylindrical specimens following the procedures of ASTM C496. The specimen was loaded through plywood strips whose width was 1/8 of the specimen diameter. The velocity of displacement of the machine actuator was 0.3 mm/min.

Stable three-point bend tests on the notched beams were carried out to determine the fracture properties of concrete following the procedures devised by Elices, Guinea and Planas [31-33]. The tests were performed in CMOD control, at a constant rate proportional to the beam depth. The rate in µm/min was 0.0667 D (D in mm) for the first 25 min, and 0.333 D to the end of the test.

Pull-out tests were carried out by pulling the wire at a constant displacement rate while keeping the concrete surface compressed against a steel plate. Figure 11 sketches the pull-out test set-up: A stiff frame fastened to the machine actuator holds the concrete specimen while the wire protrudes through a hole across the upper steel plate and is clamped to the load cell. The relative slip between the wire and the concrete surfaces at both ends of the specimen is measured by two clip gauges. The slip of the wire

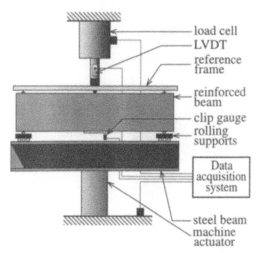

Fig. 12. Experimental set-up for the reinforced beam tests.

at the pulling side is measured relative to the stiff plate by a steel clip fastened to the wire at a point very close to the concrete surface. The tests were carried out at constant displacement rate and the load-slip curves at both ends were recorded.

Reinforced beam tests. The reinforced beams were tested in three-point bending as sketched in Fig. 12. The beam rests on two rigid-steel semi-cylinders laid on two supports permitting rotation out of the plane of the beam and rolling along the beam longitudinal axis with negligible friction. These supports roll on the upper face of a very stiff steel beam fastened to the machine actuator. The load-point displacement is measured in relation to a reference frame laid over the beam, resting on conical supports in the same vertical line as the beam supports. An LVDT inside the central loading rod measures the displacement of the central beam support relative to the frame. The test is controlled by means of a clip-gauge centered in the tensioned face of the beam, of gage length 2/3 D, and driven at a fixed rate proportional to the beam depth. The rate in μm/min was 0.0267 D (D in mm) during the first 20 min of the test, and 0.133 D up to complete failure. This kind of test control gives stable tests in which the entire post-peak behavior is recorded even for the biggest beams, in which snap-back was registered in the load-displacement curve.

Test Results

The main results for standard tests are included in Tables 1 and 2. The results of pull-out and reinforced beam tests are described in the following sections. The experimental results are compared with the curves generated by the *effective slip-length* model.

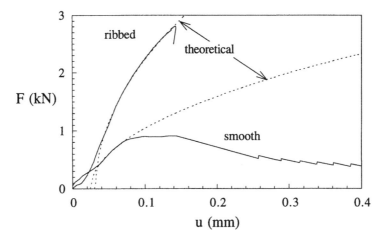

Fig. 13. Pull-out test results for smooth and ribbed wires.

Pull-Out Tests. Fig. 13 shows the load-slip curves for two typical pull-out tests (full lines) and the theoretical curves for the simplified pull-out model that gives the anchoring force in (6) —where w_s is twice the protrusion of the bar out of the concrete specimen, $w_s = 2\,u$. For the smooth wires, which slipped before breaking, τ_c was calculated as the peak force divided by the contact area. For the ribbed bars, which broke before full slip, τ_c was determined by fitting the expression in (6) to the F-u curve. In both cases an initial s-shaped portion appeared due to accommodation of the test device; this part was eliminated in the theoretical fit as shown in the figure. The agreement between the experimental curves and the simple theory is good for the ribbed wire, and only acceptable for the smooth one. The values of τ_c deduced from these tests were 0.52 ± 0.19 MPa for smooth wires and 5.3 ± 1.8 MPa for ribbed wires.

Beams with Ribbed Bars. Figure 14 shows the experimental load-displacement curves for the reinforced beams with ribbed bars (light dotted lines are the experimental curves; the experimental range is represented by the dark-grey shaded area between the experimental curves) compared to theoretical predictions given by the model (full lines). The theoretical predictions were made using the material parameters obtained from the characterization tests. The steel wires are assumed to be elastic-perfectly plastic with E_s and f_y equal to the experimental values in Table 2 (f_y is identified with the ultimate tensile strength σ_u). The concrete softening is assumed to follow the quasi-exponential function in Fig. 4b. The value of f_t was taken as the cylinder splitting strength, whose value, together with those of E_c and G_F are recorded in Table 1. The values of the bond strength used in the computations were those obtained in the pull-out tests.

The light shaded area around the theoretical prediction represents the expected standard deviation of the prediction induced by the scatter in the bond strength alone. It was determined by making various computations, keeping all the parameters equal to their corresponding mean values, except for τ_c which was varied one standard deviation above or below its mean value.

Theoretical curves obtained with higher τ_c fit the experimental curves better than those calculated with the mean value of τ_c, but the good agreement obtained between theory and experiment is remarkable since *all* the parameters used in the model were from *independent* tests.

Another remarkable result is the existence of a secondary peak when the steel cover is large, as in the 300 mm depth beams. This secondary maximum appears within the snap-back zone of the P-δ curve (Fig. 14h). The *effective slip-length* model reproduces and justifies this secondary peak: when the cover is large enough, the first maximum load is reached before the cohesive crack crosses the reinforcement, so the steel contribution to the beam resistance is very small. When the cohesive crack reaches the reinforcement, its growth is stopped by the steel, and the response of the beam is stiffer. Further crack growth produces a stretching of the reinforcement which requires a load increase (hardening). Only after the cohesive crack has grown well past the reinforcing layer does the softening of concrete dominate again the overall behavior (provided, of course, the steel ratio is small enough not to be fully dominated by the steel contribution).

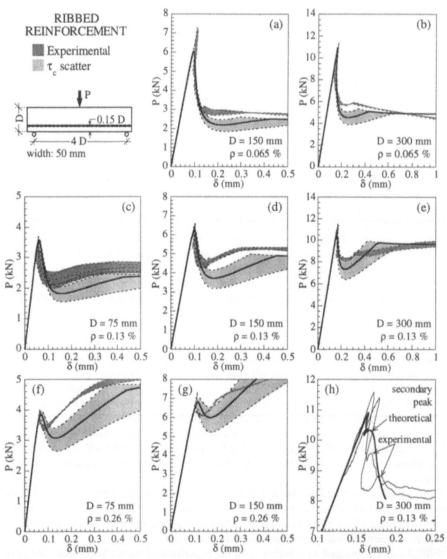

Fig. 14. (a)-(g) Experimental and theoretical P-δ curves corresponding to beams reinforced with ribbed wires. The curves in each column are of beams of the same depth, and those in each row are of beams with the same steel ratio. Light dotted lines are experimental curves; the experimental range is shown by the dark-grey shaded band. The full lines are the mean theoretical predictions (i.e. predicted for the mean values of the parameters). The light shaded band represents one standard deviation in the bond strength. (h) Zoom of the experimental and theoretical curves for D = 300 mm and ρ = 0.13%, with a secondary peak after the maximum load.

Beams with Smooth Bars. Figure 15 shows the experimental and theoretical curves of all the beams reinforced with smooth wires. The experimental curves corresponding to beams of the same type are shown as light dashed lines and their range is shaded dark grey. The *effective slip-length* model (full lines) predicts well the maximum cracking load and the ultimate yielding load, but the intermediate zone of the analytical curves is far below experimental curves. the fit is good only for D = 75 mm and ρ = 0.26%. The smooth-steel to concrete bond in the experimental beams appears above the value of τ_c measured through pull-out tests.

We calculated the τ_c value necessary to raise the theoretical curves up to the experimental ones, taking into account that τ_c must be the same for all beam depths and reinforcement ratios; $\tau_c = 0.75\, f_t$

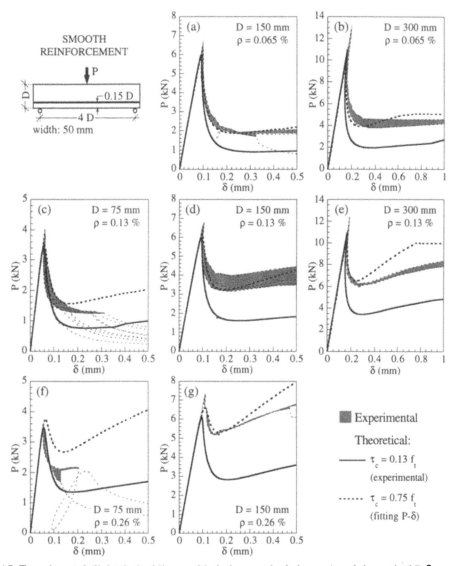

Fig. 15. Experimental (light dashed lines, with dark-grey shaded range) and theoretical P-δ curves (full line and heavy dashed line) for smooth reinforcement. The beam depth is constant in the columns, and reinforcement ratio in the rows. The dashed thick P-δ curve corresponds to $\tau_c = 0.75\, f_t$, the τ_c value with the best fitting to experimental results.

gives a reasonable fitting to the intermediate zone of the P-δ experimental curves for all sizes and steel ratios (heavy dashed lines in Fig. 15. Note that the theoretical curve for D = 150 mm and ρ = 0.13%, Fig. 15d, is within the experimental band of the P-δ curves).

Even with this correction, the predictions of the model are not fully satisfactory. One of the reasons is that for such a weak bond, the slip length is very large, so complete slip of the wire occurs before steel yielding, an effect that cannot be captured by the effective slip-length model which assumes an infinitely long beam.

DISCUSSION AND COMMENTS

Sensitivity of the Beam Tests to the Parameters of the Effective Slip-Length Model

In the present section the problem is considered of whether the reinforced beam tests are sensitive to the parameters influencing the numerical response of the effective slip-length model, i. e. size, reinforcement ratio and bond properties.

Influence of the Size. Figure 16 shows the dimensionless load-inelastic displacement curves of geometrically similar beams, reinforced with 0.13% ratio of ribbed wires with the same bond properties. In this plot the elastic displacement is subtracted from the total displacement to obtain the inelastic displacement δ_{in}, which includes the effect of cracking [34].

The mean theoretical curves (heavy lines) are surrounded by a hatched band representing the error band due to the scatter of data in the characterization testing. A 95% confidence interval was used to build the upper and lower envelopes. The experimental results for this reinforcement ratio are shown by grey bands.

The reinforced beams exhibit a mild but clear size effect: the nondimensional strength decreases as the depth increases. The theoretical maximun loads fit very well the experimental values, thus showing size effect, and the theoretical bands around the peak zone agree very well with the experimental results for each depth.

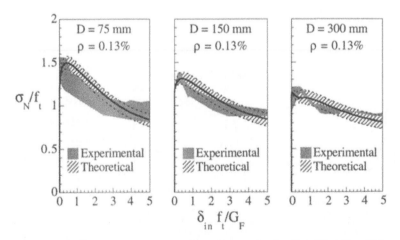

Fig. 16. Nondimensional P-δ_{in} curves for three different beam depths (D = 75, 150 y 300 mm) with the same ratio of ribbed reinforcement (ρ = 0.13%). The theoretical curves are in the middle of an error band in which each parameter varies within its 95% confidence interval. The grey bands show the experimental results.

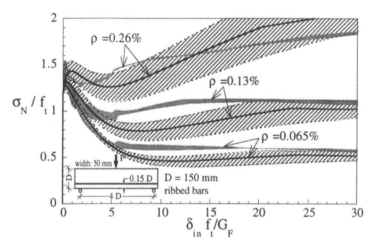

Fig. 17. Nondimensional P-δ_{in} curves of intermediate size beams (D = 150 mm) for three levels of reinforcement (ρ = 0.065, 0.13 and 0.26%). The theoretical curves are in the middle of an error band (hatched) where each parameter varies in a 95% confidence interval. The three grey bands show the experimental results.

Influence of the Reinforcement Ratio. Figure 17 shows the experimental and theoretical results of intermediate size beams (D = 150 mm) for three reinforcement ratio levels (with the same reinforcement bond properties, corresponding to ribbed wires). The results are displayed as nondimensional load-inelastic displacement curves. The curves for the theoretical model fed with the mean values of the material parameters are again surrounded by a 95% error probability band (hatched band).

Both experimental and theoretical curves are sensitive to the steel ratio. The post-peak response of the beam is stronger for higher ratios, and obviuosly the plateau corresponding to the steel yielding is strictly proportional to the reinforcement ratio since the yielding strength of the steel remained constant throughout experimentation.

The peak loads predicted theoretically show dependence on the steel ratio. The strength is higher the larger the steel ratio (a phenomenon known as hyperstrength). This effect is not clear in the experimental curves, probably because it is blurred by experimental scatter.

Influence of Bond Properties. To show the effect of the bond-slip strength on the beam behavior we focus our attention on large size beams with a 0.13% reinforcement ratio, with either ribbed or smooth wires. This size was selected because the length along which the wire is embedded within the concrete at both sides of the crack in these large beams is the longest within our experimental program (2.25 D = 675 mm) so it is the maximum slip length. In fact, smooth wires slip completely —even for the largest beams— and this deformation mechanism is not considered in the model. Nevertheless, for this long embedment length we can assume that the hypotheses of the model are fulfilled at least during the first stages of the debonding process.

Figure 18 shows the experimental and theoretical curves for these beams. In this case the hatched band around the theoretical curve for ribbed wires (Fig. 18a) represents the influence on the beam behavior of the standard deviation of τ_c around the mean value. As indicated above, the model fittings are better with values of τ_c higher than those measured; i.e., for the curves in Fig. 18 the model gives a very good approximation to the experimental curves for $\tau_c = 1.85 f_t$ (which is equal to the mean value plus the standard deviation).

The curve given by the model for the reinforced beams with weak bond bars (Fig. 18b) is far below the experimental band; this, together with the high scatter obtained for τ_c in the pull-out tests of the smooth wires, casts some doubts on the real τ_c value. In addition, feeding the model with $\tau_c = 0.75 \, f_t$ instead of with the measured mean value ($0.13 \, f_t$), the fitting to the experimental curves is quite good provided model hypotheses are accomplished, i. e., while slip length is less than half the beam length.

Although further research is required to improve the predictions of the model, it is clear that the bond-properties are essential in predicting the behavior of the beams. Both the tests and the theoretical model confirm that stronger steel-to-concrete bond leads to a stronger beam response. The model is able to describe quite accurately the behavior of lightly reinforced beams, especially for strong bond. For weak bond it might be necessary to improve the pull-out test to measure the shear bond strength τ_c.

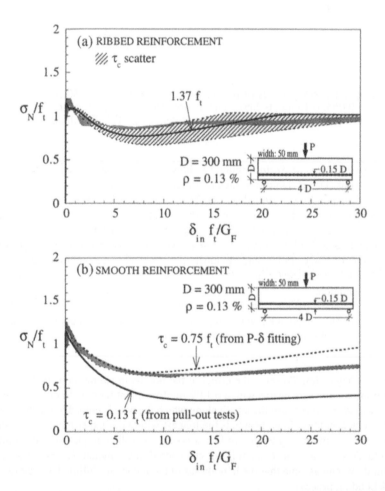

Fig. 18. Dimensionless load-inelastic displacement curves for large beams ($D = 300$ mm) and intermediate steel ratio ($\rho = 0.13\%$). (a) Ribbed wires; the grey shaded band is the experimental range, the thick full line the theoretical prediction for the mean values of the parameters, and the hatched band corresponds to the standard deviation in τ_c. (b) Smooth wires; the grey shaded band is the experimental range, the thick full line the theoretical prediction for the mean value of the parameters, and the thick dashed line the theoretical calculation for $\tau_c = 0.75 \, f_t$, which provides the best fitting for all experimental P-δ curves.

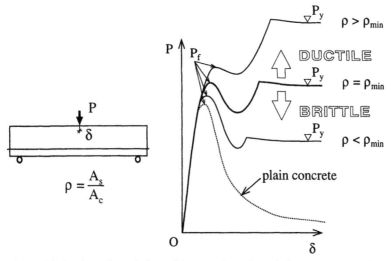

Fig. 19. Transitional behavior of a reinforced beam when the reinforcement ratio is increased. We consider that the beam is sufficiently reinforced when its behavior is in the bound between brittleness and ductility.

An Expression for Minimum Reinforcement

With regard to the necessary minimum reinforcement to ensure a ductile collapse, Fig. 19 shows the transition process of a lightly reinforced beam loaded in three-point bending when only the reinforcement ratio is varied.

The minimum steel ratio is defined as that allowing the beam a transitional behavior between brittleness and ductility, so we consider that the steel ratio is the minimum admissible if the maximum load (maximum cracking load, P_f, see Fig. 19) equals the exterior load level when the steel is working at its yield strength, and cracking is totally developed (ultimate yielding load, P_y):

$$\begin{aligned}
&\text{Brittle behavior}: & P_f &> P_y \\
&\text{Minimum steel ratio}: & P_f &= P_y \\
&\text{Ductile behavior}: & P_f &< P_y
\end{aligned} \qquad (24)$$

Outline of the Derivation of the Expression. To obtain an expression for the minimum reinforcement, ρ_{min}, using the effective slip-length model involves getting expressions for each of the sides of equation (24) and solving for the steel ratio:

$$P_f (..., \rho_{min}, ...) = P_y (..., \rho_{min}, ...) \quad \Rightarrow \quad \rho_{min} (D, c, f_y, \tau_c) \qquad (25)$$

This procedure was followed in [13], where a minimum reinforcement formula was obtained by empirical fit of numerical results, as a function of the beam depth, D, the reinforcement cover, c, the steel yielding strength, f_y, and the bond shear strength, τ_c:

$$\rho_{min} (D, c, f_y, \tau_c) = \frac{0.174}{(1 - \gamma)} \, \frac{1 + (0.85 + 2.3 \, \beta_1)^{-1}}{f_y^* - \eta_1 \, \varphi} \qquad (26)$$

where ρ_{min} is expressed as a ratio of the concrete section, and

$$\beta_1 = \frac{D}{\alpha l_{ch}} \quad , \qquad \gamma = \frac{c}{D} \quad , \qquad f_y^* = \frac{f_y}{f_t} \quad ,$$

$$\eta_1 = \left(n \frac{\tau_c}{f_t} \frac{p \, \alpha \, l_{ch}}{A_s} \right)^{1/2} \quad , \qquad \varphi = \left\langle (\beta_1)^{0.25} - 3.61 \gamma_1 \right\rangle \quad , \qquad \gamma_1 = \gamma \beta_1 = \frac{c}{\alpha l_{ch}} \tag{27}$$

$\langle \bullet \rangle$ being the Macauley brackets defined so that $\langle x \rangle = x$ if $x > 0$ and $\langle x \rangle = 0$ if $x < 0$ (φ is a nondimensional coefficient proportional to the increase in the maximum load when the steel ratio or the bond strength increase, and therefore it cannot be negative); α is a nondimensional parameter depending on the cohesive fracture behavior when a crack is initiating, and is the ratio between the fracture energies corresponding to two softening functions giving the same peak load for the beam: a linear softening and the actual softening function. It can be obtained experimentally by the procedure described in [35] and its value for an ordinary concrete is around 0.5; α can be estimated from the concrete fracture properties in the Model Code [36]. This leads to the expression:

$$\alpha = \frac{65 + 15 \, d_{max}/d_0}{170} \tag{28}$$

where d_{max} is the maximum aggregate size and $d_0 = 8$ mm.

Comparison with Other Minimum Reinforcement Formulas. To compare the foregoing formula for minimum reinforcement with those of other authors and with recommendations from building codes, we present some graphics corresponding to two types of concrete from Sections 2.1.3–4 of the CEB-FIP Model Code. The mechanical parameters are shown in Table 4: type C40 is a normal-strength and type C80 a (moderately) high-strength concrete.

For an overall picture, Fig. 20 collects the curves for minimum reinforcement to be used for beams of different depths according to various formulas. The curves are of beams of concrete C40, using a steel with a yield strength of 480 MPa and a cover of c = 40 mm (2.5 times the aggregate size). Note that contrary to the approach selected in the experimental research, the cover is assumed to be constant rather than proportional to the beam depth. Note also that the vertical axis is in logarithmic scale to keep all the curves in view and separable.

The short-dash curves shown in the figure correspond to the minimum reinforcement recommendations from the following three building codes: (1) CEB-FIP Model Code [36]; (2) ACI-318 Code [37]; and (3) Eurocode 2 [38]. The corresponding formulas are based on limit analysis and do not display size effect for geometrically similar structures. The apparent size effect in the curves is due to the fact that here non similar structures are considered, the cover being constant rather than proportional to the depth, and the minimum reinforcement in the codes refers to the effective depth D-c rather than to the total depth D.

The long-dash curves in the figure correspond to the minimum reinforcement from the formulas (based on fracture mechanics approaches) proposed by the following authors: (1) Bosco, Carpinteri and Debernardi [2] (curve labeled as 'Carpinteri', which is based on linear elastic fracture mechanics concepts and empirical results); (2) Hawkins and Hjorsetet [8] (curve labeled as 'Hawkins', which is based on finite element calculations using a cohesive crack model); (3) Balluch, Azad and Ashmawi [3] (curve labeled as 'Balluch', which is based on a modification of Bosco and Carpinteri's model [5]); and (4) Gerstle et al. [9] (curve labeled 'Gerstle', based on a simplified cohesive crack model). None of these formulas takes bond-slip into account.

Table 4. Parameters of two of the concrete-types defined in the Model Code (d_{max} = 16 mm)

Model Code concrete-type	f_{ck} MPa	f_c MPa	f_t MPa	G_F N/m	E_c GPa	l_{ch} mm	αl_{ch} mm
C40	40	48	3.5	90	36	264	148
C80	80	88	5.6	135	44	189	106

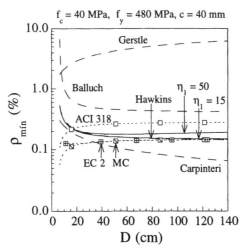

Fig. 20. Minimum reinforcement given by code specifications, by specifications of other authors and by our expression for concrete type C40, f_y = 480 MPa and c = 40 mm. Note the logarithmic scale for the vertical axis.

The full-line curves shown in the figure correspond to the minimum reinforcement as given by Eq. (26) for two bond-strength levels, characterized by parameters η_1 respectively equal to 15 and 50 which are considered practical lower and upper bounds for this characteristic (i.e., it is expected that most practical cases will fall on a curve lying between the two full-line curves).

In this figure the formula of Gerstle et al. is seen to be over-conservative; these authors defined the minimum reinforcement as that necessary to eliminate the first peak in the load-displacement curve, which leads to extremely large steel ratios. The formula by Balluch, Azad and Ashmawi also appears too conservative, although the reason is not as clear. The other curves based on fracture models show similar trends, which contrast strongly with those of the recommendations in the codes, although for beam depths larger than about 10 cm, the order of magnitude is similar. Note in particular that the minimum reinforcement based on fracture mechanics shows a strong decrease with increasing beam depth for small beam depths. This trend becomes milder for larger sizes, and for some models the minimum reinforcement becomes nearly constant for large sizes (Hawkins and Hjorsetet's and the present effective slip model with weak bond) or even slightly increasing (effective slip-length model with strong bond).

Unfortunately, this picture, although qualitatively valid, is not quantitatively universal, because the minimum reinforcement is influenced by various factors, such as the concrete quality, the steel grade and the concrete cover of the reinforcement. This is illustrated in Fig. 21, in which several cases are shown.

Figure 21a shows the case shown in Fig. 20, except that the vertical scale is now linear rather than logarithmic and the formulas of Balluch, Azad and Ashmawi and of Gerstle et al. are not shown. Fig. 21b shows a similar plot in which only the concrete quality has been changed (concrete C80 in Table 4 is used). Note that the code requirements are unchanged (except for Eurocode 2 for beam depths less than 60 cm), while all the formulas based on fracture mechanics lead to an increase of the minimum reinforcement, with trends qualitatively similar to those in Fig. 21a. The reason is that the behavior of high strength concrete is more brittle, a fact that is well known but requires the use of fracture mechanics to be adequately incorporated into the models and formulas.

Figure 21c shows a plot similar to that in Fig. 21a in which the yield strength of the reinforcement is halved. As a rule of thumb, the minimum reinforcement should be doubled, to keep the product ρf_y constant. This is seen to be essentially so with the minimum reinforcement recommended by the codes (note that the scale of the vertical axis in Fig. 21c has been modified so that the curves would look

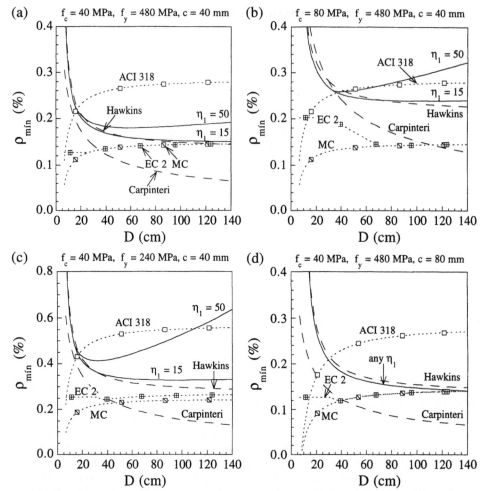

Fig. 21. Minimum reinforcement predicted by various formulas. Full lines are the predictions according to Eq. (26). (a) Reference case. (b) Effect of increasing the concrete grade. (c) Effect of decreasing (halving) the steel yield strength. (d) Effect of increasing (doubling) the cover.

identical to those in Fig. 21a were the product ρf_y kept constant). The same is true of the formulas of Bosco, Carpinteri and Debernardi, and of Hawkins and Hjorsetet. A very small effect of reducing the reinforcement yield strength appears in the formula derived from the effective slip-length model if the bond strength is low, but the effect becomes important when the bond strength is high.

The effect of the bond strength in the minimum reinforcement appears in Eq. (26) in the denominator of the second fraction as a term $\eta_1\varphi$ which is subtracted from the dimensionless yield strength f_y^*. Only when $\eta_1\varphi$ is comparable to f_y^*, does the bond-strength appreciably affect the minimum reinforcement.

Figure 21d shows a plot similar to that in Fig. 21a in which the only variation is the cover, which is doubled (it is made equal to 5 times the aggregate size). We note the following effects: (1) the curve of Bosco, Carpinteri and Debernardi is insensitive to the cover, (2) the curves from the codes require less reinforcement (due to the decrease in the effective beam depth for a given total beam depth), (3) the formulas of Hawkins and Hjorsetet and Eq. (26) for low bond strength require more reinforcement, especially for small beam depths, and (4) the effect of bond strength disappears. This is because, for the range of sizes under consideration, the cover is so large that the first peak of the load occurs before the cohesive zone reaches the reinforcement, and so neither the presence of the reinforcement or the bond strength affect the first peak load. Note also that for this case the minimum reinforcement required according to the formula of Hawkins and Hjorsetet and according to Eq. (26) are almost coincident.

SUMMARY

1. Lightly reinforced beams, which fail through a single crack in the middle cross section, can be analyzed using fracture mechanics models, in particular the cohesive crack model was used for this research. The bond between steel and concrete is modelled by the effective slip-length model, which is amenable to computationally efficient numerical treatment, and helps to disclose the main parameters influencing the behavior of lightly reinforced beams.

2. An experimental research was conducted on lightly reinforced beams of reduced size. The properties of the microconcrete are such that the observed behavior is representative of beams of ordinary size made of ordinary concrete. The experiments studied the effect of steel ratio, beam depth and bond strength, and included the determination of all the parameters of the model by independent tests.

3. The numerical predictions of the experimental results using the effective slip-length model are reasonably good, considering that the model includes simplifications and that no parameter optimization is allowed (all the parameters were determined from independent tests). The model is able to capture minute experimental details, such as a secondary peak in the load for relatively large steel covers, but requires better estimates of the bond between steel and concrete than those obtained from pull-out tests. Better accuracy can no doubt be achieved by implementing more sophisticated descriptions of the material behavior into the model (e.g. by including strain-hardening in the stress-strain curve of steel, using a nonlinear bond-slip relationship instead of a rigid-plastic model, and using more accurate softening curves for concrete) and, particularly, by improving the measurement of the bond strength, which displayed a very large scatter in the pull-out tests in this research.

4. The numerical model captures adequately the transitional behavior of lightly reinforced beams from brittleness to ductility as the reinforcement ratio is increased and, in this respect, closely matches the experimental results. The numerical model was used to generate a closed-form expression for minimum reinforcement in bending which is compared to recommendations from building codes and formulas from other authors. The comparison shows that the recommendations in the codes could be improved to get safer or cheaper minimum reinforcement to avoid brittle behavior. It also shows that the influence of bond strength should be taken into account, and that, generally speaking, larger reinforcement is required the larger the bond strength, although the quantitative effect of the bond strength depends on the details of the beam, as regards the geometry (reinforcement cover) and the materials (concrete and steel grade).

ACKNOWLEDGEMENTS

The authors gratefully acknowledge financial support from Comisión Interministerial de Ciencia y Tecnología, CICYT Spain, under Grants MAT 97-1022 and MAT 97-1007-C02-2.

REFERENCES

1. Bosco, C., Carpinteri, A., and Debernardi, P.G. (1990) "Fracture of reinforced concrete: Scale effect and snap-back instability". *Engineering Fracture Mechanics*, **35**(4-5), 665-667.
2. Bosco, C., Carpinteri, A., and Debernardi, P.G. (1990) "Minimum reinforcement in high-strength concrete". *Journal of Structural Engineering* (ASCE), **116**(2), 427-437.
3. Baluch, M.H., Azad, A.K., and Ashmawi, W. (1992) "Fracture mechanics application to reinforced concrete members in flexure". In: *Applications of Fracture Mechanics to Reinforced Concrete*, A. Carpinteri (Ed.). Elsevier Applied Science, London, 413-436.
4. Hededal, O., and Kroon, I.B. (1991) "Lightly reinforced high-strength concrete". M. Sc. Thesis. Åalborg University, Denmark.
5. Bosco, C., and Carpinteri, A. (1992) "Fracture mechanics evaluation of minimum reinforcement in concrete structure". In: *Applications of Fracture Mechanics to Reinforced Concrete*, A. Carpinteri (Ed.) Elsevier Applied Science, London, 347-377.
6. Bosco, C., and Carpinteri, A. (1992) "Fracture behavior of beam cracked across reinforcement". *Theor. Appl. Fract. Mech.*, **17**, 61-68.
7. Massabò, R. (1994) "Meccanismi di rottura nei materiali fibrorinforzati". Ph.D. Thesis. Ingegneria Strutturale. Politecnico di Torino, Italy.
8. Hawkins, N.M., and Hjorsetet, K. (1992) "Minimum reinforcement requirements for concrete flexural members". In: *Applications of Fracture Mechanics to Reinforced Concrete*, A. Carpinteri (Ed.) Elsevier Applied Science, London, 379-412.
9. Gerstle, W.H., Dey, P.P., Prasad, N.N.V., Rahulkumar, P., and Xie, M. (1992) "Crack growth in flexural members—A fracture mechanics approach". *ACI Structural Journal*, **89**(6), 617-625.
10. Ulfkjær, J.P., Hededal, O., Kroon, I., and Brincker, R. (1994) "Simple application of fictitious crack model in reinforced concrete beams—Analysis and experiments". In: *Size Effect in Concrete Structures*, H. Mihashi, H. Okamura and Z.P. Bazant (Eds.), E & FN Spon, London, 281-292.
11. Ruiz, G., Planas, J. and Elices, M. (1993) "Propagación de una fisura cohesiva en vigas de hormigón debilmente armadas". *Anales de Mecánica de la Fractura*, **10**, 141-146.
12. Ruiz, G., and Planas, J. (1994) "Propagación de una fisura cohesiva en vigas de hormigón débilmente armadas: modelo de la longitud efectiva de anclaje". *Anales de Mecánica de la Fractura*, **11**, 506-513.
13. Ruiz, G. (1996) "El efecto de escala en vigas de hormigón débilmente armadas y su repercusión en los criterios de proyecto". Ph.D. Thesis. Dpto. de Ciencia de Materiales, E.T.S. de Ingenieros de Caminos, Canales y Puertos, Universidad Politécnica de Madrid, Spain.
14. Planas, J., Ruiz, G., and Elices, M. (1995) "Fracture of lightly reinforced concrete beams: theory and experiments". In: *Fracture Mechanics of Concrete and Concrete Structures*, F.H. Wittmann (Ed.), Aedificatio Publishers, Freiburg, 1179-1188.
15. Hillerborg, A., Modeer, M., and Petersson, P.E. (1976) "Analysis of crack formation and crack growth in concrete by means of fracture mechanics and finite elements", *Cement and Concrete Research*, **6**, 773-782.
16. Planas, J. and Elices, M. (1986) Un nuevo método de análisis del comportamiento de una fisura cohesiva en Modo I, *Anales de Mecánica de la Fractura*, **3**, 219-227.
17. Planas, J. and Elices, M. (1992) Asymptotic analysis of a cohesive crack: 1. Theoretical background, *Int. Journal of Fracture*, **55**, 153-177.
18. Bazant, Z.P. (1990) Smeared-tip superposition method for nonlinear and time-dependent fracture, *Mech. Res. Commun.*, **17**(5), 343-351.
19. Bazant, Z.P. and Beissel, S. (1994) Smeared-tip superposition method for cohesive fracture with rate effect and creep, *Int. J. Fracture*, **65**, 277-290.

20. Planas, J., and Elices, M. (1992) "Shrinkage eigenstresses and structural size effect". In: *Fracture Mechanics of Concrete Structures*, Z. P. Bazant (Ed.), Elsevier Applied Science, London, 939-950.

21. Planas, J., and Elices, M. (1993) "Drying shrinkage effect on the modulus of rupture". In: *Creep and Srinkage of Concrete* , Z. P. Bazant and I. Carol (Eds.), E & FN Spon, London 357-368.

22. Bazant, Z. P., and Sener, S. (1988) "Size effects in pull-out tests". *ACI Materials Journal*, **85**, 347-351.

23. Malvar, L. J. (1992) "Confinement stress dependent bond behavior. Part I: experimental investigation". In: *Bond in Concrete. From Research to Practice*, Riga Technical University and CEB (Eds.), Vol. 1, 79-88.

24. Mazars, J., Pijaudier-Cabot, G., and Clement, J. L. (1992) "Analysis of steel-concrete bond with damage mechanics: non linear behaviour and size effect". In: *Applications of Fracture Mechanics to Reinforced Concrete*, A. Carpinteri (Ed.), Elsevier Applied Science, London, 307-331.

25. Morita, S., Fuji, S., and Kondo, G. (1994) "Experimental study on size effect in concrete structures". In: *Size Effect in Concrete Structures*, H. Mihashi, H. Okamura and Z. P. Bazant (Eds.), E & FN Spon, London, 21-40.

26. ACI Committee 446, Fracture Mechanics (1992) "State-of-art report". In: *Fracture Mechanics of Concrete Structures*, Z. P. Bazant (Ed.), Elsevier Applied Science, London, 1-140.

27. Bazant, Z. P., and Cedolin, L. (1980) "Fracture mechanics of reinforced concrete". *J. of the Eng. Mech. Div.*, ASCE, **106** (EM6), 1257-1306.

28. Petersson, P. E. (1981) "Crack growth and development of fracture zones in plain concrete and similar materials". Report TVBM-1006. Division of Building Materials, Lund Institute of Technology, University of Lund, Sweden.

29. Bache, H. H. (1994) "Design for ductility". In *Concrete Technology: New Trends, Industrial Applications*, A. Aguado, R. Gettu, and S. P. Shah (Eds.), E & FN Spon, London, 113-125.

30. Planas, J., and Elices, M. (1991) "Nonlinear fracture of cohesive materials". *International Journal of Fracture*, **51**, 139-157.

31. Guinea, G. V., Planas, J. and Elices, M. (1992) "Measurement of the Fracture Energy using Three Point Bend Tests. 1. Influence of experimental procedures". *Materials and Structures*, **25**, 121-218.

32. Planas, J., Elices, M. and Guinea, G. V. (1992) "Measurement of the Fracture Energy using Three Point Bend Tests. 2. Influence of bulk energy dissipation". *Materials and Structures*, **25**, 305-312.

33. Elices, M., Guinea, G. V., and Planas, J. (1992) "Measurement of the Fracture Energy using Three Point Bend Tests. 3. Influence of cutting the P-δ tail". *Materials and Structures*, **25**, 327-334.

34. Planas, J., Guinea, G. V., and Elices, M. (1992) "Stiffness associated with quasi-concentrated loads". *Materials and Structures*, **27**, 311-318.

35. Planas, J., Guinea, G. V., and Elices, M. (1994) "SF-2. Draft test method for linear initial portion of the softening curve of concrete". In: *Draft Proposal to the Committee of the JCI International Collaboration Project on Size Effect in Concrete Structures*. Dpto. de Ciencia de Materiales, E.T.S. de Ingenieros de Caminos, Canales y Puertos, Universidad Politécnica de Madrid, Spain.

36. CEB-FIP Model Code (1991) "CEB-FIP Model Code 1990, Final Draft". In: *Bulletin D'Information*, N. **203**, **204** & **205**, EFP Lausanne.

37. ACI Committee 318 (1993) "Building Code requirements for reinforced concrete (ACI 318-89) (Revised 1992) and commentary—ACI 318-R89 (Revised 1992)". In: *ACI Manual of Concrete Practice*, Part **3**. American Concrete Institute, Detroit.

38. Eurocode 2, Subcommittee CEN/TC 250/SC2, (1993) "Eurocode 2: Design of concrete structures (ENV 1992-1-1:1991)". Commission of the European Communities (Ed.), Paris.

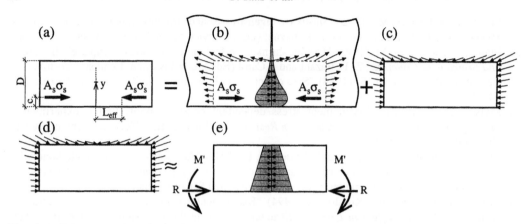

Fig. A.1. Scheme used to obtain the stresses generated by the reinforcement along the central cross section of the beam.

APPENDIX A: INTERNAL STRESSES CAUSED BY THE REINFORCEMENT

The stresses generated by the closing forces F_s at the central cross section of an uncracked elastic beam (Fig. A.1a) can be calculated by superposing the stresses they would generate in an elastic halfspace (Fig. A.1b) and those corresponding to the beam (Fig. A.1c). The first case was solved by Melan [A.1]; the second is handled approximately by assuming that it produces a linear stress distribution on the central cross-section which preserves overall equilibrium of each half of the beam. (i.e. the surface tractions are reduced to a force and a bending moment on the central cross section) as indicated in Fig. A.1e.

The stress distribution corresponding to the case in Fig. A.1b (the halfspace) can be written as

$$\bar\sigma_1 = \frac{A_s\,\sigma_s}{B}\,F(y,\,c,\,L_{eff}) \tag{A.1}$$

where y, c, and L_{eff} are the dimensions defined in Fig. A.1a, and F is the stress distribution for a unit closing force (per unit thickness), derived by Melan as

$$F = \frac{L_{eff}}{2\,\pi(1-v)}\left[\frac{L_{eff}^2}{r_1^4} + \frac{L_{eff}^2 + 4cy - 2c^2}{r_2^4} + \frac{8cyL_{eff}^2}{r_2^6} + \frac{1-2v}{2}\left(\frac{1}{r_1^2} + \frac{3}{r_2^2} - \frac{4y\,(c+y)}{r_2^4}\right)\right] \tag{A.2}$$

where

$$r_1^2 = (y-c)^2 + L_{eff}^2 \quad \text{and} \quad r_2^2 = (y+c)^2 + L_{eff}^2 \tag{A.3}$$

The stress distribution (A.1) must be $A_s\,\sigma_s$, and thus the function $F(y,\,c,\,L_{eff})$ satisfies:

$$\int_0^\infty F\,(y,\,c,\,L_{eff})\,dy = 1 \tag{A.4}$$

Likewise, the moment of the stress distribution relative to a point on the lower free surface must be $c\,A_s\,\sigma_s$ and therefore $F(y,\,c,\,L_{eff})$ also satisfies the condition:

$$\int_0^\infty y\,F\,(y,\,c,\,L_{eff})\,dy = c \tag{A.5}$$

To get the overall stress distribution, the foregoing distribution (A.1) must be corrected to include the stresses in Fig. A.1e. For this correction, we calculate the resultant R and the bending moment M' as

$$R = A_s \sigma_s R^* \quad \text{and} \quad M' = A_s \sigma_s D M'^* \ , \tag{A.6}$$

where

$$R^* = \int_D^\infty F(y, c, L_{eff}) \, dy = \left[1 - \int_0^D F(y, c, L_{eff}) \, dy \right] \tag{A.7}$$

$$M'^* = \frac{1}{D} \int_D^\infty y \, F(y, c, L_{eff}) \, dy = \frac{1}{D} \left[c - \int_0^D y \, F(y, c, L_{eff}) \, dy \right] \tag{A.8}$$

and then the stress distribution in Fig. A.1e is readily found to be

$$\bar{\sigma}_2 = \sigma_1 + \sigma_2 \frac{y}{D} \tag{A.9}$$

where

$$\sigma_1 = \frac{2 A_s \sigma_s}{BD} (R^* - 3M'^*) \quad \text{and} \quad \sigma_2 = \frac{6 A_s \sigma_s}{B} (2 M'^* - R) \tag{A.10}$$

from which it follows that, adding the correction (A.9) to the distribution (A.1), the overall stress distribution at the central cross section is given by

$$\bar{\sigma} = \bar{\sigma}_1 + \bar{\sigma}_2 = \frac{A_s \sigma_s}{BD} \left[D \, F(y, c, L_{eff}) + 2 \, (R^* - 3M'^*) + 6 \, (2M'^* - R^*) \frac{y}{D} \right] \tag{A.11}$$

The expression in brackets in (A.11) is a nondimensional function dependent on D, y, c and L_{eff}, because R^* and M'^* in (A.7) and (A.8) also depend on those variables. So Eq. (A.11) can be written as:

$$\bar{\sigma} = \rho \, \sigma_s \, F'(D, y, c, L_{eff}) \tag{A.12}$$

where ρ is the steel ratio ($\rho = A_s/BD$) and $F'(\cdot)$ is the expression in brackets in (A.11).

Appendix A Reference

A.1 Melan, E., (1932) "Der spannungzustand der durch eine einzelkraft im innern beanspruchten halbschiebe". *2 Angew. Math. Mech.* **12**, Band 12, Heft 6, 343-346

BEHAVIOUR OF R/C ELEMENTS IN BENDING AND TENSION: THE PROBLEM OF MINIMUM REINFORCEMENT RATIO

A.P. FANTILLI[1], D. FERRETTI[2], I. IORI[2] and P. VALLINI[1]

[1]*Department of Structural Engineering, Politecnico di Torino, 10129 Torino, Italy*
[2]*Department of Civil Engineering, University of Parma, 43100 Parma, Italy*

ABSTRACT

To study the first cracked stage in R/C members in tension or bending two monodimensional models are proposed. The models analyse the transition from the pre-cracked stage to the post-cracked one by assuming a bond-slip relationship and one cohesive crack. The accuracy of these assumptions is checked by comparing the numerical results with some experimental data. The models are used to compute the minimum reinforcement ratio and to enlighten its size-effect.

KEYWORDS

R/C beams, R/C tendons, bond-slip, nonlinear fracture mechanics, minimum reinforcement, size effect.

INTRODUCTION

The problem of minimum reinforcement ratio in flexural or tensile reinforced concrete (R/C) members cannot be disjoined from a deep analysis of their physical behaviour. In particular, when the crack starts and grows, two different physical aspects are involved: the bond-slip between reinforcing steel and concrete, and the fracture mechanics of concrete. However, with the aim at obtaining straightforward formulas for the minimum reinforcement ratio, several simplified hypotheses of the physical reality are usually introduced. A classical hypothesis assumes perfect bond between the reinforcing steel and the surrounding concrete, by ignoring the possible slippage and the corresponding bond stresses between the two materials. As known, this hypothesis could lead to a rough definition of deformability in the first cracked stage, where the minimum reinforcement ratio is computed. This stage is usually modelled only by considering fracture mechanics, obtaining an "asymptotic model". In other words, the approach of fracture mechanics alone is able to enlighten a final trend but neglects the transitory phase that precedes this final "asymptote". By means of more detailed models, it is possible to describe this transitory phase and to quantify with more accuracy the "asymptote". Detailed models are obtained by observing the experimental transitory phase, usually called softening branch. Experimental evidence shows that the softening branch is strongly affected both by bond properties of the adopted reinforcing bar (i.e. bar diameter, area and geometry of its surface) and by the stress field nearby the crack. Therefore, if we took into account only the fracture mechanics in the cracked cross section, we could partially reproduce the behaviour of some R/C members. In other words, by means of a metaphor, when R/C claims its rights, we concede it only some help for its natural survival. Therefore, the aforementioned remarks should be rewarded for the minimum reinforcement ratio problem.

In the following sections, the minimum reinforcement ratio is computed for tensile and flexural members by introducing a unified approach which is able to join fracture mechanics and bond behaviour. For sake of simplicity, this unified approach is initially proposed to model the softening branch of tensile members. Once its validity is experimentally checked, the model is used to compute the minimum reinforcement ratio. Subsequently, the same steps are repeated for flexural members by introducing suitable enhancements and extensions of the model.

BEHAVIOUR OF R/C ELEMENTS IN TENSION

Some General Remarks

Extensive researches have been carried out to study the behaviour of reinforced concrete tensile members. Since tensile members involve two major aspects of R/C, that is concrete fracture and bond between steel and concrete, they allow starting the study of R/C in a simple but not oversimplified way. In particular, they allow enlightening in a rational way several problems of R/C structures like strength, deformability and cracking (e.g. crack width, spacing and pattern). Early studies on tensile members were focused on the determination of the crack width, crack spacing and deformability, with special regard to the so-called "stabilised crack pattern". The results obtained for R/C tensile members have been extended to flexural members and are nowadays the basis of some important code rules [1, 2]. In particular, deformability has been widely studied in tensile members both from an experimental and from a theoretical point of view, with special regard to the so-called "tension-stiffening".

Since the first studies of Mörsch in [3], it is well-known that a block of concrete included between two consecutive cracks is able to support tensile stresses. This phenomenon increases the stiffness of the bar by a quantity known as tension-stiffening. Bond behaviour is an essential aspect of tension-stiffening since it controls the ability of the reinforcement to transfer tensile stresses to concrete. Therefore, theoretical studies have been carried out by removing the hypothesis of perfect bond between steel and concrete and by allowing a slip between the two materials. A bond-slip relationship has been used to model numerically uniaxial tensile members by means of bi-dimensional and mono-dimensional approaches. Bi-dimensional approaches, such as finite element method [4, 5, 6] and boundary element method [7], model concrete and reinforcing bars by means of appropriate elements. By using finite element method, the two materials are linked by introducing bond-link elements or contact elements that are able to simulate the assumed bond-slip relationship. Mono-dimensional approaches suppose that concrete cross sections remain plane so as to avoid the study of the behaviour of the whole cross section [8, 9]. The above numerical models allow to study contemporaneously both crack width and crack pattern as well as tension stiffening, by giving suitable average stress-strain relationships σ_s - ε_{sm}.

Recently, there has been a renewed interest in the characterisation of the stress-strain behaviour of tensile members, probably due to the increasing use of theories which employ in their formulations average constitutive relationships of the cracked concrete [2, 10, 11]. These models require average stress-strain relationships at any given loading stage, from the initiation of the first crack up to the final stabilized crack pattern. In the early cracked stage, crack width is very small and it seems necessary to model the crack formation by means of a discrete cohesive crack model. Moreover, the early cracked stage is useful to study low reinforced members where the cracking load might be close to the ultimate load.

Proposed Model

Aim of this section is to propose a numerical model that is able to describe the mechanical behaviour of R/C tensile members during the formation of the first crack (Fig. 1a). In this stage, the load-elongation diagram of the tendon can reveal a stable behaviour (strain softening), as well as an unstable one (snap-back) [12]. The unstable behaviour can be numerically and experimentally observed by controlling the crack width (Fig. 1a), which is a monotonically increasing variable. This is the same as to impose, in the cracked cross section, a slip $s = w/2$ between the reinforcing bar and the surrounding concrete. The effect of such a slip influences the tensile member within a distance $z = l_{tr}$ (known as transfer length) from the cracked section (Fig. 1b).

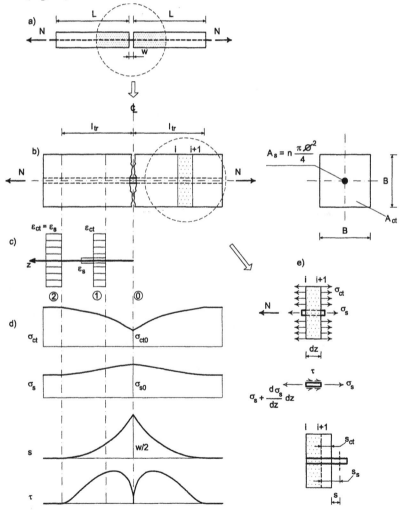

Fig. 1. Tensile member with one crack: a) geometry; b) transfer length l_{tr}; c) strain profiles; d) distribution of concrete stress σ_{ct}, steel stress σ_s, slip s and bond stress τ; e) free body diagrams.

Within the transfer length l_{tr} , the tensile stresses diffuse from the steel to the surrounding concrete (Fig. 1d) by means of bond stresses τ. Conversely, outside the zone of diffusion, perfect bond between steel and concrete is assumed. In particular, according to Stage I hypotheses, the strains of reinforcing bar ε_s and concrete ε_c are equal and strain profiles are assumed to be plane. Within the transfer length, on the contrary, the strains of the two materials are different but the concrete strain profile is still assumed to be plane (Fig. 1c). Although some tests point out the presence of greater strains near the reinforcing bar [13, 14], the hypothesis of plane strain profile seems acceptable according to [8, 9,15].

For a generic cross section of the tendon, together with the compatibility assumptions shown in Fig. 1c, the equilibrium equation is considered, namely (Fig. 1e):

$$\int_{A_{ct}} \sigma_{ct}\, dA + \int_{A_s} \sigma_s\, dA = N \tag{1}$$

where σ_{ct} and σ_s are the stresses in concrete and steel respectively, A_{ct} and A_s are the areas of concrete and steel and N is the applied tensile force. For a portion of the tendon with infinitesimal length dz the equilibrium equation of the reinforcing bar can be written as (Fig. 1e):

$$\frac{d\sigma_s}{dz} = \frac{p_s}{A_s} \cdot \tau(s(z)) \tag{2}$$

where p_s is the perimeter of reinforcement, and $\tau(s(z))$ is the bond stress. For the same portion of the tendon, it is possible to define the slip $s(z)$ as the difference of the displacements between two initially overlapping points belonging to steel and concrete (Fig. 1e):

$$s(z) = s_s(z) - s_{ct}(z) \tag{3}$$

Therefore, by derivation with respect to z, Eq. (3) becomes:

$$\frac{ds}{dz} = -[\varepsilon_s(z) - \varepsilon_{ct}(z)] \tag{4}$$

where ε_s is the strain of the reinforcing bar and ε_{ct} is the tensile strain in the surrounding concrete.

To solve the problem, the equilibrium and compatibility equations must be associated with the constitutive laws of the two materials and with the bond-slip relationship τ-s. In particular, the concrete in tension is assumed to have a linear elastic stress-strain relationship, with modulus of elasticity E_c (Fig. 2a).

Initial tension crack starts to form when the tensile strength of concrete f_{ct} is reached in a specific weak section. The crack growth is modelled by means of the fictitious crack model proposed in [1] and represented by the following relationships (Fig. 2b):

$$
\begin{aligned}
\sigma_{ct} &= E_c \cdot \varepsilon_{ct} && , \textit{if } w = 0 \\[2mm]
\sigma_{ct} &= 0.15 \cdot f_{ct} + \frac{w_1 - w}{w_1} \cdot 0.85 \cdot f_{ct} && , \textit{if } 0 \le w < w_1 \\[2mm]
\sigma_{ct} &= 0.15 \cdot f_{ct} \cdot \frac{w_c - w}{w_c - w_1} && , \textit{if } w_1 \le w \le w_c \\[2mm]
\sigma_{ct} &= 0 && , \textit{if } w > w_c
\end{aligned}
\tag{5}
$$

For reinforcing steel (Fig. 2c) the following stress-strain relationship is adopted:

$$\sigma_s = E_s \cdot \varepsilon_s \quad , if \ \varepsilon_s \leq \varepsilon_{sy} \tag{6}$$

Again in accordance with [1], the bond-slip relationship (Fig. 2d) is:

$$
\begin{aligned}
\tau &= \tau_{max} \cdot \left(\frac{s}{s_1}\right)^\alpha && , if \ 0 \leq s < s_1 \\
\tau &= \tau_{max} && , if \ s_1 \leq s < s_2 \\
\tau &= \tau_{max} - (\tau_{max} - \tau_f) \cdot \frac{(s - s_2)}{(s_3 - s_2)} && , if \ s_2 \leq s < s_3 \\
\tau &= \tau_f && , if \ s \geq s_3
\end{aligned}
\tag{7}
$$

The parameters given in Fig. 2d are valid for ribbed reinforcing steel and depend on the confinement, bond conditions and concrete strength. Nearby the cracks, the effect of debonding reduces the bond stress by a factor λ:

$$\lambda = \frac{z}{\varnothing/5} \leq 1 \tag{8}$$

where z represents the distance from the crack and $\varnothing$ is the bar diameter.

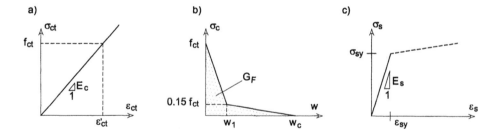

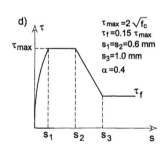

	unconfined concrete		confined concrete	
value	bond conditions		bond conditions	
	good	all other cases	good	all other cases
s_1	0.6 mm	0.6 mm	1.0 mm	1.0 mm
s_2	0.6 mm	0.6 mm	3.0 mm	3.0 mm
s_3	1.0 mm	2.5 mm	clear rib spacing	
α	0.4	0.4	0.4	0.4
τ_{max}	$2.0\sqrt{f_{ck}}$	$1.0\sqrt{f_{ck}}$	$2.5\sqrt{f_{ck}}$	$1.25\sqrt{f_{ck}}$
τ_f	$0.15\,\tau_{max}$	$0.15\,\tau_{max}$	$0.40\,\tau_{max}$	$0.40\,\tau_{max}$

Fig. 2. Constitutive laws [1]: a) uncracked concrete in tension; b) cracked concrete in tension; c) reinforcing steel; d) bond-slip.

To solve the problem, in addition to the foregoing equations, suitable boundary conditions are also adopted. In particular, for the cracked cross section ($z = 0$) cohesive stress conditions are introduced:

$$s_0 = w/2$$
$$\sigma_{ct,0} = \sigma_{ct}(w) \tag{9}$$
$$\sigma_{s,0} = N - A_{ct} \cdot \sigma_{ct,0}$$

For the section with perfect bond ($z = l_{tr}$), Stage I conditions are required:

$$s(z = l_{tr}) = 0$$
$$\sigma_{ct}(z = l_{tr}) = \frac{N}{A_{ct} + \dfrac{E_s}{E_c} \cdot A_s} \tag{10}$$
$$\sigma_s(z = l_{tr}) = \frac{E_s}{E_c} \cdot \sigma_{ct}(z = l_{tr})$$

From a mathematical viewpoint, Eqs. (1 - 8) and boundary conditions (9-10) constitute a differential problem called "two point boundary value problem". In principle this mathematical problem can be solved with the aid of various numerical techniques [16], such as relaxation through finite differences [9], finite element method [17] and shooting technique [8]. The "shooting method" is adopted herein for the numerical solution. In particular the solution is found by starting from the cracked cross section where the slip is equal to $w/2$ and therefore the stress $\sigma_{ct,0}$ in concrete is known. Conversely, a trial value for the applied load N is assumed. In this way, Eqs. (1 - 8) can be integrated from the cracked cross section up to the opposite state I cross section, where the boundary Eqs. (9 - 10) are checked. If these requirements are not satisfied, the value of the load N must be adjusted by means of a trial and error procedure explained by the following algorithm:

1. Select a crack width w
2. Assume a value of the load N
3. Compute $\sigma_{ct,0}$, $\sigma_{s,0}$ according to Eqs. (9)
4. Put $i = i + 1$ and $z_i = z_{i-1} + \Delta z$
   Integrate Eqs. (2,4) by means of the implicit multistep method of Adams Moulton triggered with Adams Bashforth [16], to obtain $\sigma_{ct,i}$, $\sigma_{s,i}$, s_i
5. If $s_i > 0$ then go to step 4 else go to step 6
6. By assuming $l_{tr} = z_i$ and $m = i$, if boundary Eqs. (10) are not verified, then assume a new trial value of the load N and go back to step 3.

The solution of the problem yields the distribution along the block of all the static and kinematic unknowns (Fig. 1d): the stress in concrete (σ_{ct}) and steel (σ_s), the slip (s) and the bond stress (τ). Consequently, the elongation ΔL of the tensile member can be determined as:

$$\Delta l = \int_L \varepsilon_s \, dz = \sum_{i=1}^{m} \varepsilon_{s,i} \cdot \Delta z \tag{11}$$

where L is the length of the considered reference portion (Fig. 1a).

Some Numerical Results

The accuracy of the proposed model is checked by comparing (Fig. 3) the computed stresses and strains with the experimental results by Scott & Gill [14]. In particular, Fig. 3a shows a good agreement between the computed steel strains and the values measured for the specimen 100T12 (Table 1). Therefore, the elongation ΔL can be computed correctly by means of Eq. (11), despite the simplified bond-slip relationship adopted. With this relationship, the computed bond stresses can partially show the variability of the experimental values (Fig. 3b), due to all the mechanical aspects involved in bond behaviour (splitting cracks, debonding, micro-cracking, confinement, etc.).

However, the proposed model is able to simulate correctly the main physical aspects such as the N-ΔL curve, so that it can be used to predict phenomena of practical interest. In particular, the model is used to simulate the behaviour of the tensile members shown in Fig. 4. According to [1], by assuming the maximum aggregate size equal to 16 mm, the computed value of the fracture energy is $G_F = 64.1$ N/m.

Table 1. Properties of tested tensile members

Specimen Ref.	B [mm]	f_c [N/mm²]	E_s [N/mm²]	n	∅ [mm]	E_c [kN/mm²]	G_F [N/m]
40x40 [18]	40	29.6	200000	1	12	28.0	-
100T12 [14]	102	37.4	200000	1	12	33.5	42.7

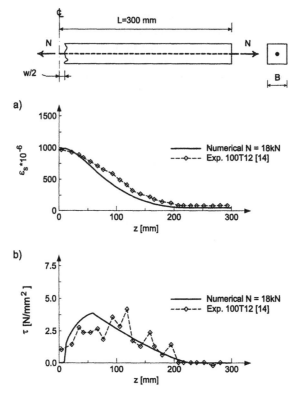

Fig. 3. Comparison between numerical outcomes and experimental data [14]: a) steel strains ε_s; b) bond stresses τ.

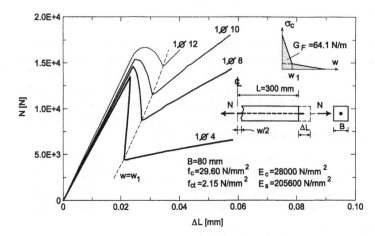

Fig. 4. Tension-elongation N-ΔL curves with various diameters $\varnothing$ of reinforcing bar.

To enlighten the effect of reinforcement, four tendons with different bar diameter (4, 8, 10 and 12 mm) are examined. The proposed numerical model supplies the tension vs. elongation N-ΔL diagrams presented in Fig. 4.

When the crack starts ($w < w_1$), two opposite contributions dictates the structural response: the reduction of cohesive stresses in concrete and the increase of stresses in reinforcing bar. If the bond contribution is small (small bar diameter $\varnothing$ and area A_s), the cohesive contribution prevails and a significant softening branch of the diagram N-ΔL appears. This softening branch tends to reduce as the diameter of the reinforcing bar increases. When the crack grows ($w > w_1$), the cohesive contribution virtually vanishes and the numerical outcomes show an abrupt change in the structural response which is dictated by the reinforcing bar. For this reason, if a small bar is used, the structural response does not reach the peak tension antecedent the softening branch and the failure is brittle.

Successively, to enlighten the effect of cohesion, the tensile member reinforced with 1 $\varnothing$ 12 is analysed with two different values of the fracture energy G_F (42.7 N/m and 64.1 N/m). The outcomes, depicted in Fig. 5, show that a diminishing of the fracture energy reduces the peak tension and modifies the subsequent softening branch from stable to unstable (according to the definition proposed in [12, 19]).

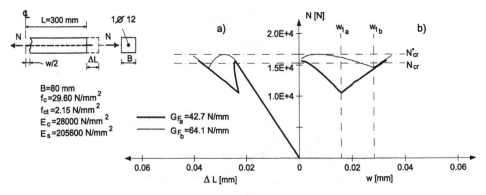

Fig. 5.   Importance of the fracture energy G_F: a) tension-elongation N-ΔL curves; b) tension-crack width N-w curves.

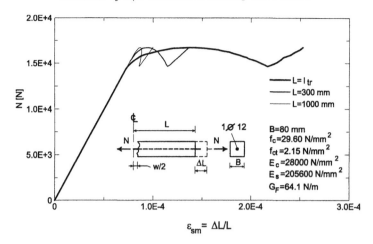

Fig. 6. Tension-average steel strain N-ε_{sm} curves with various reference lengths L.

The structural response can also be expressed as a function of the average strain ε_{sm} defined as:

$$\varepsilon_{sm} = \frac{\Delta L}{L} \tag{12}$$

where L represents the length of the examined portion. The diagrams N-ε_{sm} shown in Fig. 6 reveal that, by increasing the length L of the examined portion, the structural response becomes brittle.

In particular, if the reference portion L is longer than l_{tr}, the average strain ε_{sm} is affected by the Stage I behaviour of the portion exceeding l_{tr}. In this portion the diminishing of the load N (as w increases) causes an elastic unloading that could transform the response of the whole tensile member from a softening one (when $L = l_{tr}$) to a snap-back one (when $L > l_{tr}$) [12]. Therefore, to follow experimentally the snap-back branch, it is necessary to test the specimens by locating and controlling w, which is a monotonically increasing parameter (Fig. 5b). Sometimes, the softening branch shown in Fig. 5 is preceded by a peak load (effective crack load $N^*{}_{cr}$) larger than the theoretical load at first crack N_{cr}:

$$N_{cr} = f_{ct} \cdot \left(A_{ct} + \frac{E_s}{E_c} \cdot A_s \right) \tag{13}$$

If the applied force N exceeds N_{cr}, the concrete tensile stress reaches and rises above the tensile strength f_{ct} in some cross sections of the tendon, paradoxically without the formation of a new crack. This anomaly can be accepted by supposing that the first crack develops in a specific weak section in which the f_{ct} corresponds to the minimum value among the natural scatter of the tensile strength of the member. When the natural scatter is small, nearby cracks could form, as confirmed experimentally by Gijsbers and Hehemann in [13] and Scott & Gill in [14] for uniaxially tensioned members.

For loads N grater than $N^*{}_{cr}$, Stage I condition is possible only by admitting values of the tensile strength more and more far from the minimum value f_{ct} of the weak section. In this situation, the hypothesis of single crack is less reliable and the formation of several cracks is taken into account [8]. In particular the entire tensile member is modelled through a succession of blocks divided by cracks. The behaviour of each block is governed by the aforementioned compatibility and equilibrium Eqs. (1 - 8) together with the cohesive boundary Eqs. (9) for the two cracked edges of each block [8].

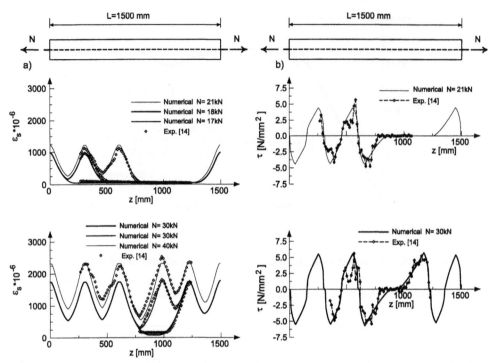

Fig. 7.   Comparison between numerical outcomes and experimental data [14] when many cracks occur:
a) steel strains; b) bond stresses.

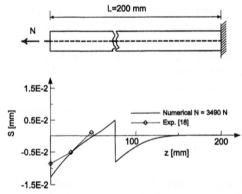

Fig. 8.   Comparison between numerical slips and experimental values measured by means of speckle
photography technique [18].

In this way, it is possible to follow the evolution of the crack pattern for the tensile member already
shown in Fig. 3. The steel strains within the tendon, measured by means of strain gauges, show a high
scatter in the first crack load N_{cr}. Due to this scatter, numerical cracks are introduced according to both
experimental crack loads and positions (Fig. 7) [14].

The analysis is repeated for the tensile member 40x40 (Table 1) tested by Iori and Rosati in [18]. For
this member, Fig. 8 shows the comparison between the slips computed with the proposed model and
measured by means of the laser speckle photography technique.

The Minimum Reinforcement Ratio

For tensile members with a single crack, Fig. 4 shows different structural responses obtained by varying the reinforcement ratio $\rho = A_s/A_{ct}$. In particular, with a small value of ρ (corresponding to 1 Ø 4 in Fig. 4), the tension load N reached after the crack formation is always smaller than the effective crack load N^*_{cr} and the failure is brittle. Therefore, to prevent this sudden failure a minimum amount of reinforcement is required. Referring to Fig. 9, the minimum reinforcement area $A_{s,min}$ can be defined as the minimum area that fulfils the condition:

$$N_y = N^*_{cr} \qquad (14)$$

where N^*_{cr} represents the effective crack load and N_y stands for the tension that produces the yielding of reinforcement (namely $\sigma_{s,max} = \sigma_{sy}$).

The minimum reinforcement ratio ρ_{min} strongly depends on the effective crack load N^*_{cr} which can be augmented by increasing the fracture energy G_F (Fig. 5) or by increasing Ø and A_s consequently (Fig. 4). The effective crack load N^*_{cr} can also be augmented by incrementing the perimeter of the reinforcement (Fig. 10a) by using more bars of smaller diameter and keeping constant their area ($A_s = 78.5$ mm^2). When this situation occurs, the softening branch is less evident, as depicted in Fig. 10b. For this reason ρ_{min} should be defined by considering both geometrical and mechanical properties (fracture energy, tensile strength and structural size) and bar diameter.

The values of ρ_{min} are depicted in Fig. 11 for tensile members with different dimensions B of the side and two diameters Ø of the reinforcing bar.

The two horizontal lines represented in Fig. 11a demonstrate that the minimum reinforcement ratio is affected by the diameter Ø but is independent on the dimension B. This result can be explained by observing Fig. 11. In particular the tensile member of Fig. 11b is reinforced by using one bar of diameter Ø corresponding to ρ_{min}. If the side B is doubled (Fig. 11c) or triplicated (Fig. 11d), the behaviour of the first tensile member is repeated and the minimum reinforcement ratio remains the same. In other words, in R/C tendons the minimum reinforcement ratio is not affected by a structural size effect on B if the mechanical properties of the materials remain unchanged.

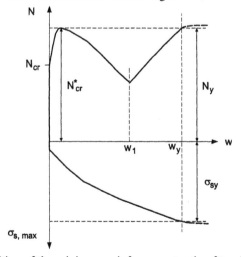

Fig. 9. Definition of the minimum reinforcement ratio of tensile members.

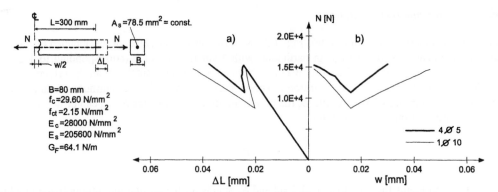

Fig. 10. Analysis of tensile members with the same area and different perimeter of reinforcing bars: a) load-elongation *N*-Δ*L* curves; b) load-crack width *N*-*w* curves.

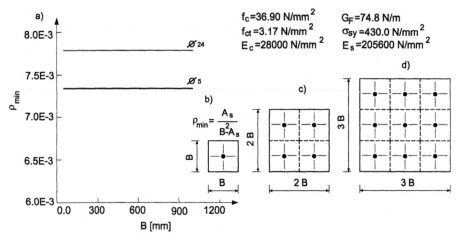

Fig. 11. Minimum reinforcement ratio and its structural size effect.

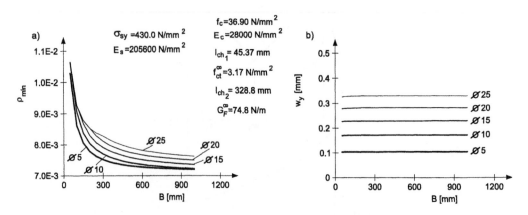

Fig. 12. Material size effect by means of the multifractal scaling law [22, 23]: a) minimum reinforcement ratio ρ_{min}; b) crack width w_y at yielding.

Nevertheless, it is well known that the nominal tensile strength f_{ct} and fracture energy G_F display a size effect. In particular f_{ct} is constant for relatively large sizes, whereas it increases for relatively small sizes [20]. It is possible to take into account these considerations to compute the minimum reinforcement ratio ρ_{min} by means of an appropriate size effect law for the mechanical properties of concrete [21]. The tensile strength and the fracture energy can be expressed by using the multifractal scaling law of [22] and [23] respectively:

$$f_{ct}(B) = f_{ct}^{\infty} \cdot \left[1 + \frac{l_{ch,1}}{B}\right]^{1/2} \tag{15}$$

$$G_F(B) = G_F^{\infty} \cdot \left[1 + \frac{l_{ch,2}}{B}\right]^{-1/2} \tag{16}$$

where:

B = linear size of the cross section (Fig. 1a);

l_{ch1}, l_{ch2} = internal characteristic lengths;

f_{ct}^{∞} = value of f_{ct} obtained for $B \to \infty$;

G_F^{∞} = value of G_F obtained for $B \to \infty$.

By using a concrete mix with the same mechanical characteristics reported in [24], new values of the minimum reinforcement ratio are computed.

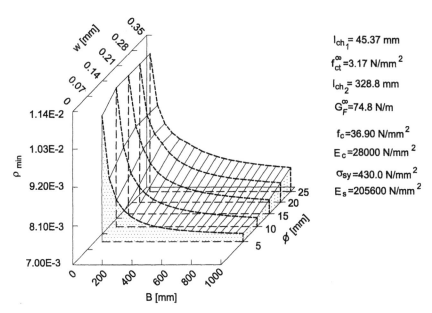

l_{ch_1}= 45.37 mm

f_{ct}^{∞}=3.17 N/mm^2

l_{ch_2}= 328.8 mm

G_F^{∞}=74.8 N/m

f_c=36.90 N/mm^2

E_c=28000 N/mm^2

σ_{sy}=430.0 N/mm^2

E_s=205600 N/mm^2

Fig. 13. Minimum reinforcement ratio ρ_{min} and crack width w_y by varying B and $\varnothing$.

In particular, Fig. 12a shows a clear reduction of ρ_{min} as the size B increases. According to [21, 25], this reduction is only due to a size effect on material properties. Moreover, the minimum reinforcement is required not only to provide a ductile response and to ensure adequate warning of an incipient failure at overloads, but also to prevent excessive crack width at service load. Therefore, for the previous condition of minimum reinforcement (Fig. 12a), the relationship existing between the size B and the crack width at yielding w_y is represented in Fig. 12b.

The same figure shows that the crack width w_y is practically independent of the dimension B while it changes by employing different diameters $\varnothing$ of the bar.The minimum reinforcement ratio ρ_{min} and the corresponding crack width w_y can be depicted in a three-dimensional diagram as a function of B and $\varnothing$ (Fig. 13). Since w_y is practically independent of B, the crack width and the diameter $\varnothing$ are represented along the same axis.

From a practical point of view, the diagram could be used to evaluate $\varnothing$ and ρ_{min} by starting from the size B and the allowable crack width w_y.

BEHAVIOUR OF R/C ELEMENTS IN BENDING

Some General Remarks

The considerations made in the previous section are repeated and extended in the present section for beams in bending whose deformability is usually defined by means of the moment-curvature relationship M-$1/r$. Due to cracking phenomenon, in R/C beams it is difficult to define only one cross-sectional moment-curvature relationship. In fact, for the same bending moment M it is possible to obtain different curvatures $1/r$ for cracked or uncracked cross sections. For this reason, for a representative portion of the beam, an average curvature comprised between Stage I and Stage II curvatures is usually considered (Fig. 14). This M-$1/r$ relationship has the same meaning of the N-ε_{sm} law previously introduced for tensile R/C members. Furthermore, like in tensile members, the transition from the uncracked Stage I to the stabilized crack pattern (first cracked stage) claims for particular attention. In this situation, a simplified change of deformability as the linear joint between the pre-cracked stage and the post-cracked one (dashed line in Fig. 14), could be an oversimplification [26]. In fact, when the first crack forms, both the tests carried out by controlling the displacement [27] and the crack mouth opening w [28, 29] show a remarkable softening behaviour (Fig. 14).

As pointed out by Giuriani and Rosati in [28], the study of the softening branch, which is particularly important in lightly-reinforced concrete beams, requires the knowledge of the local behaviour near the

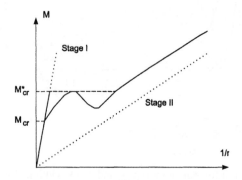

Fig. 14. Moment-curvature M-$1/r$ relationships.

crack. In particular, in the early cracked stage, crack width is very small and it seems necessary to take into account the bond-slip behaviour between reinforcing steel and concrete, as well as non-linear fracture mechanics of tensile concrete. Since the first studies developed by means of fracture mechanics, the local deformability near a crack is usually considered by assuming perfect bond between steel and concrete. Therefore, the deformability of the beam is attributed only to fracture mechanisms in the process zone which can be studied according to [30] by using a cohesive model [1, 31], or by means of LEFM [32] or by considering both a cohesive model and LEFM [33].

The models based on the aforementioned hypotheses provide consistent results only when perfect bond occurs. This is true, for example, when the crack affects only the concrete cover. On the contrary, when the crack goes through the reinforcing zone, slip between concrete and bars occurs in a certain portion of the beam, named "transfer length l_{tr}". Within the transfer length, the variation of stresses and strains

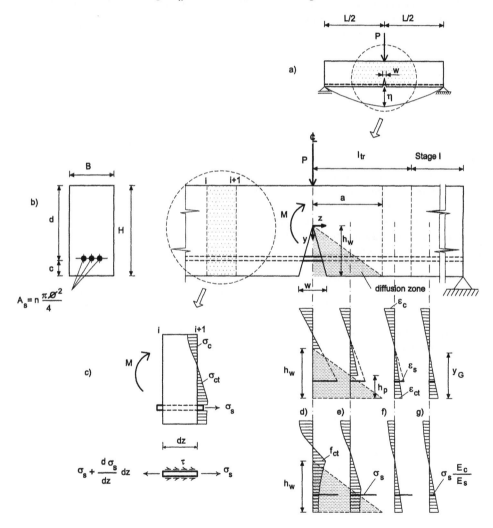

Fig. 15. Three point bending beam with one crack: a) geometry; b) transfer length l_{tr}; c) free body diagrams; d)-g) strain and stress profiles.

in steel and concrete due to bond stresses should be considered. In the literature, the bond-slip behaviour is included [29, 34] only by using an experimental pull-out relationship coupled with a cohesive crack model. Unfortunately the pull-out relationship is not general as a constitutive law and no information is provided within the transfer length.

Proposed Model

Purpose of this section is to propose a numerical model for R/C beams in bending able to compute the stress and strain fields within the transfer length l_{tr} during the formation and growth of the first crack.

The flexural deformability of R/C beams with a single crack (Fig. 15a) could be studied by means of two- or three-dimensional finite element models [35]. Nevertheless, as already proposed for tensile members, the problem can be solved by means of a one-dimensional model by introducing a suitable simplified strain profile for every cross-section of the beam.

In particular, the beam of Fig. 15a is divided in two different portions. Within the first portion, of length l_{tr}, the tensile stresses of the reinforcing bar are transmitted to the surrounding concrete by means of bond stresses τ. Conversely, within the second portion, perfect bond occurs and Stage I condition applies (Fig. 15g).

In the midspan cracked cross section (Fig. 15d), the strain profile for steel and uncracked concrete is assumed to be plane. Moreover, in the same cross-section the crack width w is assumed to vary linearly from the crack mouth ($y = h_w$) to the crack tip ($y = 0$). In the fracture process zone, cohesive stresses are introduced by means of a fictitious crack model. According to the model proposed for plain concrete beams by Giuriani and Rosati [36], the cohesive stresses propagate in the diffusion zone depicted with dotted hatch in Fig. 15b. In a generic cross-section within the diffusion zone (Fig. 15e), plane strain profile is assumed for steel and concrete in compression, whereas the strain of tensile concrete above the diffusion zone lays on a different plane. In the diffusion part h_p of the same cross-section, only the equilibrium condition is considered.

Finally, for a cross-section comprised between diffusion and Stage I zones, two different planes represent the strain profile (Fig. 15f). Although aiming to a simplified representation of physical behaviour, this assumptions find confirmation both in laboratory tests [37, 38] and numerical analyses [6, 39].

For a generic section of the block, together with these compatibility assumptions, the equilibrium equations are considered, namely:

$$\int_{A_c} \sigma_c dA_c + \sigma_s \cdot A_s = 0 \tag{17}$$

$$\int_{A_c} \sigma_c \cdot y_c \, dA_c + \sigma_s \cdot y_s \cdot A_s = M(z) \tag{18}$$

where σ_c and σ_s are the stresses in concrete and steel respectively, A_c and A_s are the areas of concrete and steel and $M(z)$ is the applied bending moment (Fig. 15c). For a portion of the beam of infinitesimal length dz, the equilibrium equation for the reinforcing bar alone can be written:

$$\frac{d\sigma_s}{dz} = \frac{p_s}{A_s} \cdot \tau(s(z)) \tag{19}$$

where p_s is the perimeter of the reinforcement in tension, and $\tau(s(z))$ is the bond stress. Referring to the same beam portion, it is also possible to define the slip $s(z)=s_s(z)-s_{ct}(z)$ as the difference in the displacements between two initially overlapping points belonging to steel and concrete. Hence, by derivation with respect to z:

$$\frac{ds}{dz} = -[\varepsilon_s(z) - \varepsilon_{ct}(z)] \tag{20}$$

where ε_s is the strain of the reinforcing bar and ε_{ct} is the tensile strain in the surrounding concrete. To solve the problem, the equilibrium and compatibility equations must be associated with the constitutive laws of the materials and with the bond-slip relationship τ-s. In particular, the ascending branch (Fig. 16a) of the relationship proposed in [1] is considered as reference for concrete in compression:

$$\sigma_c = \frac{E_c/E_{c1} \cdot \varepsilon_c/\varepsilon_{c1} - (\varepsilon_c/\varepsilon_{c1})^2}{1 + (E_c/E_{c1} - 2) \cdot \varepsilon_c/\varepsilon_{c1}} \cdot f_c \tag{21}$$

Concrete in tension is assumed to have a linear elastic stress-strain relationship, with modulus of elasticity E_c (Fig. 16b). Initial tension crack starts to form when the tensile strength of concrete f_{ct} is reached in the midspan section of the beam. The crack is modelled by means of the fictitious crack model represented by the relationships proposed in [31] and shown in Fig. 16c:

$$
\begin{aligned}
\sigma_{ct} &= f_{ct} & &, if \ w \leq w_1 \\[1mm]
\sigma_{ct} &= f_{ct} - 0.7 \cdot f_{ct} \cdot \frac{w - w_1}{w_2 - w_1} & &, if \ w_1 < w \leq w_2 \\[1mm]
\sigma_{ct} &= 0.3 \cdot f_{ct} \cdot \frac{w_3 - w}{w_3 - w_2} & &, if \ w_2 < w \leq w_3
\end{aligned}
\tag{22}
$$

For the reinforcing steel in tension or compression, the following bilinear law is assumed (Fig. 16d):

$$\sigma_s = E_{s1} \cdot \varepsilon_s \quad , if \ \varepsilon_s \leq \varepsilon_{sy} \tag{23}$$

The bond-slip τ-s relationship is adopted according to [1]:

$$
\begin{aligned}
\tau &= \tau_{max} \cdot \left(\frac{s}{s_1}\right)^\alpha & &, if \ 0 \leq s < s_1 \\[1mm]
\tau &= \tau_{max} & &, if \ s_1 \leq s < s_2 \\[1mm]
\tau &= \tau_{max} - (\tau_{max} - \tau_f) \cdot \frac{(s - s_2)}{(s_3 - s_2)} & &, if \ s_2 \leq s < s_3 \\[1mm]
\tau &= \tau_f & &, if \ s \geq s_3
\end{aligned}
\tag{24}
$$

with the suggested reduction of bond stresses near cracks (Fig. 16e).

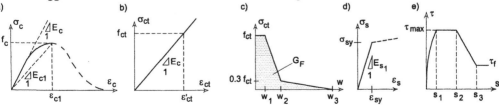

Fig. 16. Constitutive laws: a) concrete in compression [1]; b) uncracked concrete in tension; c) cracked concrete in tension [31]; d) reinforcing steel; e) bond-slip [1].

For the solution of the problem, suitable boundary conditions are required. In particular, for the cracked cross section ($z = 0$) cohesive conditions apply:

$$s_0 = \frac{w}{2} \cdot \left(\frac{h_w - c}{h_w} \right)$$

$$\sigma_{ct,0}(y) = \sigma_{ct}(w(y)) \qquad\qquad , 0 \le y \le h_w$$

$$\int_{A_c} \sigma_c dA_c + \sigma_{s,0} \cdot A_s = 0 \tag{25}$$

$$M_0 = \int_{A_c} \sigma_c \cdot y_c \, dA_c + \sigma_{s,0} \cdot y_s \cdot A_s$$

For the opposite cross-section ($z = l_{tr}$), where perfect bond occurs, Stage I conditions apply:

$$s_{ltr} = 0$$

$$\sigma_{ct,ltr}(y) = \frac{M(z = ltr)}{J_I} \cdot y' \qquad\qquad , 0 \le y' \le y_G \tag{26}$$

$$\sigma_{s,ltr} = \frac{E_s}{E_c} \cdot \frac{M(z = ltr)}{J_I} \cdot (y_G - c)$$

where J_I is the Stage I moment of inertia.

From a mathematical viewpoint, Eqs. (17 - 24), together with the boundary conditions (25 - 26), constitute a differential "two point boundary value problem". Like in tensile members, this problem can be solved with the aid of various numerical techniques. In particular, the same "shooting method" is adopted herein for the numerical solution [40].

Starting from the cracked cross section, where the crack mouth opening w is controlled, a trial value of h_w is assumed. In this fashion, like in tensile members, Eqs. (17 - 24) can be integrated from the cracked cross section up the opposite cross-section, where perfect bond occurs. The value of h_w is adjusted till all the boundary conditions are verified, by means of the following trial and error procedure:

1. Select a crack width w
2. Assume a value of the crack length h_w
3. Compute $\sigma_{ct,0}$, $\sigma_{s,0}$, M_0, according to Eqs. (25)
4. Put $i = i + 1$ and $z_i = z_{i-1} + \Delta z$
   Integrate Eqs. (19, 20) by means of the implicit multistep method of Adams Moulton triggered with Adams Bashforth [16] to obtain $\sigma_{ct,i}$, $\sigma_{s,i}$, s_i
5. If $s_i > 0$ then go to step 4 else go to step 6
6. By assuming $l_{tr} = z_i$ and $m = i$, if Stage I Eqs. (26) are not verified then assume a new trial value of the load h_w and go back to step 3.

For an imposed value of w, the solution of the problem yields the state of tensile stress in concrete and steel (σ_{ct}, σ_s), the slip (s) and the bond stress (τ) along the beam (Fig. 17). As the state of stress and the consequent state of strain are known, it is possible to compute the rotation ϕ of a beam portion l:

$$\phi = \int_l \frac{\varepsilon_s - \varepsilon_c}{d} \cdot dz = \sum_{i=1}^{m} \frac{\varepsilon_{s,i} - \varepsilon_{c,i}}{d} \cdot \Delta z. \tag{27}$$

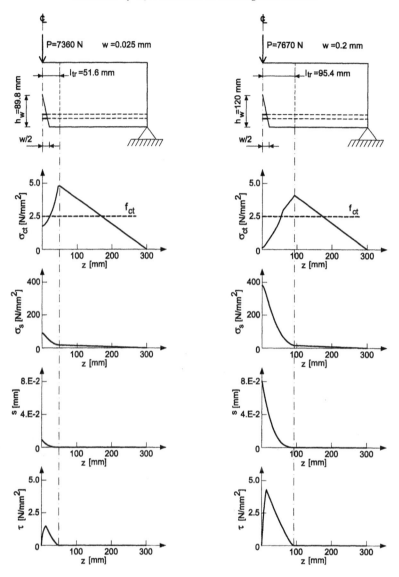

Fig. 17. Numerical states of stress (σ_{ct}, σ_s), slip (s) and bond stress (τ) for the beam D2-4XR (Table 2).

Some Numerical Results

The validity of the assumptions introduced in the proposed model is checked by modelling a beam tested by Giuriani and Rosati in [28] by means of a powerful full field displacements technique: the geometric moiré (Fig. 18b). The mechanical and geometrical properties of this beam are summarized in Table 2. In Fig. 18 the numerical moment-rotation M-ϕ diagram and the experimental one are depicted, carried out controlling the CMOD. The proposed model seems capable to simulate as the remarkable softening branch of the M-ϕ diagram (Fig. 18a) as the evolution of the crack length h_w (Fig. 18c) and the crack with w (Fig. 18d). Furthermore, the numerical model seems also capable to compute correctly the load-displacement P-η curve during the crack growth. In fact, for a statically determinate beam, it is possible to compute, for a given w, the corresponding bending moments and curvatures. Therefore the applied load P is obtained by means of equilibrium equations and the displacement η is computed by integrating curvatures. This procedure is applied to the three point bending beams tested by Planas et al. in [29] and Bosco et al. in [27], whose mechanical and geometrical properties are shown in Table 2. In the same table the related brittleness numbers N_p [41] are also reported:

$$N_p = \frac{f_y \cdot H^{0.5} \cdot A_s}{K_{Ic} \cdot B \cdot H} \tag{28}$$

where K_{Ic} is the fracture toughness of concrete.

Table 2. Properties of tested beams in bending

Beam Ref.	B [mm]	H [mm]	f_c [N/mm^2]	σ_{sy} [N/mm^2]	n	$\varnothing$ [mm]	E_c [kN/mm^2]	N_p	ρ ‰
Ø6 [28]	50	250	38.1	481	1	6	29.2	-	2.26
B1[27]	150	200	77.5	569	1	5	34.3	0.13	0.65
A3[27]	150	100	77.5	441	2	8	34.3	0.73	6.70
D2-2XR [29]	50	150	39.5	538	2	2.5	30.3	0.20	1.30
D2-4XR [29]	50	150	39.5	538	4	2.5	30.3	0.40	2.60

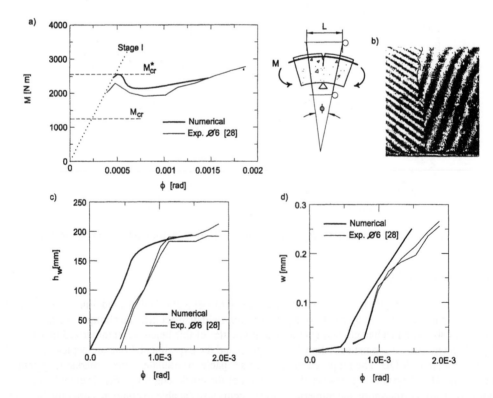

Fig. 18. Comparison between numerical outcomes and experimental data [28]: a) moment-rotation M-ϕ; b) testing equipment; c) crack height- rotation h_w-ϕ; d) crack width-rotation w-ϕ.

Figure 19 shows a substantial agreement between numerical and experimental outcomes, both for the beam B1 [27], which has a low value of the brittleness number N_p, and for the beam A3 [27], which has an higher value of N_p. The same agreement still remains by reducing the dimensions of the beams, as shown in Fig. 19 for the members D2-2XR and D2-4XR tested by Planas et al. in [29]. Therefore, during the formation of the first crack, the numerical model is able to simulate the stable (high N_p) or unstable (low N_p) transition in R/C beams of different size. Like in tensile members, the first crack starts when the tensile strength f_{ct} is reached in the tension surface. The corresponding cracking moment M_{cr} (Fig. 18a) can be computed by using Stage I hypotheses:

$$M_{cr} = f_{ct} \cdot W_I \tag{29}$$

where W_I is the section modulus of the beam.

When the crack grows, the bending moment can increase and reach a peak value called effective cracking moment M^*_{cr} [42]. Starting from this moment the structural behaviour strongly varies and shows in most cases a remarkable softening branch. Some code rules [1, 2, 43] suggest to evaluate the effective cracking moment M^*_{cr} with Eq. (29) by using a flexural tensile strength $f_{ct,fl}$ greater than f_{ct}. For this reason, when the bending moment in the cracked cross-section is greater than M_{cr}, the numerical outcomes sometimes can show (Fig. 17) wide portion of the beam where the tensile stress σ_{ct} in the tension surface exceeds f_{ct}. This phenomenon is more evident in lightly R/C beams with high bending moment gradients and good bond properties. Therefore, by considering one crack only, the proposed numerical model partially violates the experimental evidence. For lightly R/C beams with only one crack, Rokugo et al. in [44] detected, by means of strain gauges applied on the tension surface, many peaks in the concrete strains. These peaks, according to [44], are due to additional invisible microcracks nearby the main

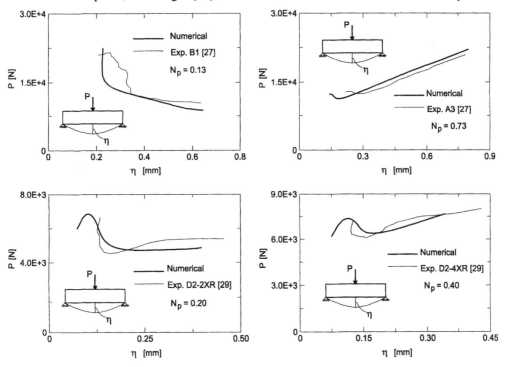

Fig. 19. Comparison between numerical load-deflection P-η curves and experimental data by changing the brittleness number N_p (Table 2).

macrocrack. The microcrack affects the development of slips and tensile stresses σ_{ct} along the beam, as shown qualitatively in Fig. 20a.

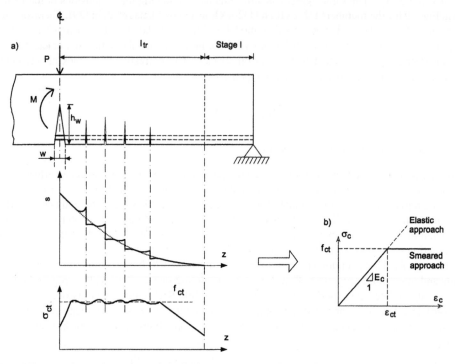

Fig. 20. Microcracks and macrocrack in R/C beam: a) experimental evidence [44]; b) elastic-plastic model for smeared crack approach.

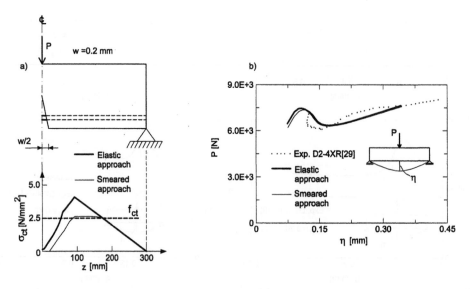

Fig. 21. Comparison between the elastic and the smeared crack approaches for the beam D2-4XR [29]: a) concrete stress σ_{ct} on the tensile surface; b) load deflection P-η curves.

When some cracks occur, the beams should be considered as a sequence of blocks delimited by two consecutive cracks. Therefore, as shown in [40], the behaviour of each block is again represented by Eqs. (17 - 24), although new boundary Eqs. (25) apply. This approach might be cumbersome for lightly R/C beams with many microcracks and only one macrocrack. As the tensile stress σ_{ct} is almost constant in the microcracked zone ($\sigma_{ct} = f_{ct}$), Fig. 20a suggests a straightforward approach. In particular, it seems useful to adopt the previous model for one macrocrack (Eqs. (17 - 24)) and smear out the microcracks by means of an elastic-plastic stress-strain relationship of tensile concrete (Fig. 20 b).

The new model is used to simulate the structural behaviour of the beam D2-4XR [29], which exhibits small transfer length and thus violates the tensile strength (Fig. 17). Compared to the previous model, the smeared approach furnishes a different stress distribution (Fig. 21a) although almost the same load deflection diagram (Fig. 21b).

The Minimum Reinforcement Ratio

The experimental P-η curves obtained by controlling the displacement η and the crack mouth opening w show various structural responses (Fig. 19). In fact, in some situations, the trend of the P-η curves monotonically increases while, in other situations, the first ascendent branch is followed by a remarkable softening. In lightly R/C beams (e.g. B1 and D2-2XR in Fig. 19) the yielding of reinforcement may occur for a bending moment lower than M^*_{cr} and the failure is called "brittle". On the contrary, in beams with a higher reinforcement ratio $\rho = A_s/BH$, after the softening branch hardening appears with $M_y > M^*_{cr}$ and the failure is called "ductile" (e.g. A3 and D2-4XR in Fig. 19). Like in tensile members, to prevent this brittle failure a minimum amount of reinforcement is required (Fig. 22). Referring to the same figure, the minimum reinforcement area $A_{s,min}$ can be defined as the minimum steel area that fulfils the condition:

$$M_y = M^*_{cr} \tag{30}$$

where M_y stands for the bending moment that produces the yielding of reinforcement (namely $\sigma_{s,max} = \sigma_{sy}$).

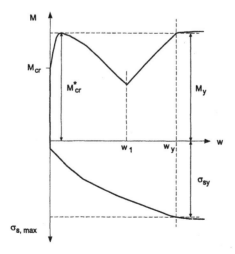

Fig. 22. Definition of the minimum reinforcement ratio for beams in bending.

At the yielding of reinforcing bar, crack width is huge and cohesive stresses can be neglected to compute the yielding moment M_y. Therefore, from Eq. (30), the minimum amount of reinforcement area $A_{s,min}$ can be expressed as:

$$A_{s,min} = \frac{M^*_{cr}}{\chi \cdot H \cdot \sigma_{sy}} \tag{31}$$

where χH stands for the flexural lever arm of the internal couple.

Hence, the minimum reinforcement ratio is given as:

$$\rho_{min} = \frac{A_{s,min}}{B \cdot H} = \frac{M^*_{cr}}{\chi \cdot B \cdot H^2 \cdot \sigma_{sy}} \tag{32}$$

It is important to enlighten the parameters that affect the effective crack moment M^*_{cr} from which ρ_{min} depends proportionally.

Both tests and numerical simulations show the dependence of M^*_{cr} on the same parameters affecting N^*_{cr} in tension. In particular, if A_s increases, the crack width is narrow and the cohesive stresses produce greater M^*_{cr}. The increases of cohesive stresses and M^*_{cr} are also obtained by augmenting the fracture energy G_F or by increasing the bond stresses (e.g. by using for the same area A_s more bars of smaller diameter).Therefore, as shown in Fig. 23, the minimum reinforcement ratio depends on the adopted bar diameter. The same figure shows a structural size effect on ρ_{min} due to the section height H.

Conversely this size effect does not appear by increasing the cross section base B (Fig. 24). Like in tensile members (Fig. 11), the behaviour of the first beam reinforced with ρ_{min} (Fig. 24b) is repeated when the base B is twice (Fig. 24c) or three times (Fig. 24d).

Finally, if a material size effect on f_{ct} is introduced by means of the multifractal scaling law, Eq. (15), the minimum reinforcement ratio ρ_{min} reduces by increasing the height H (Fig. 25a).

Differently from tensile members (Fig. 12b), the crack width at yielding depends on the size H of the beam (Fig. 25b). From a practical point of view, the curves in Fig. 25 can be used, for a given size H, to obtain the bar diameter necessary to fulfill the allowable crack width w_y.

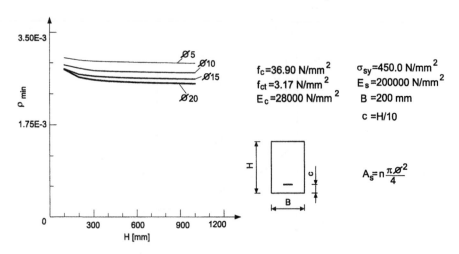

Fig. 23. Influence of the section height H and the bar diameter $\varnothing$ on ρ_{min}.

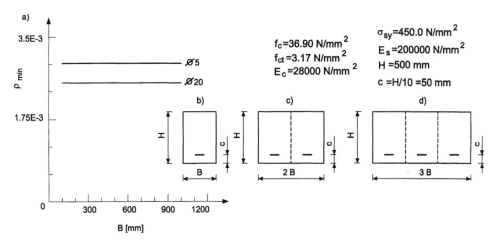

Fig. 24 Influence of the section base B and the bar diameter $\varnothing$ on ρ_{min}.

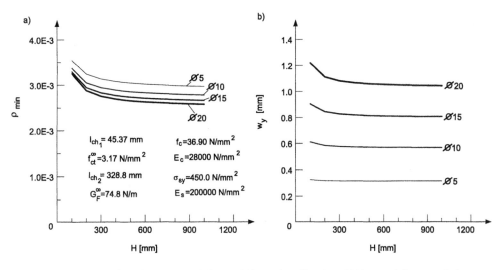

Fig. 25. Material size effect by means of the multifractal scaling law [23]: a) minimum reinforcement ratio ρ_{min}; b) crack width at yielding w_y.

CONCLUSIONS

Based on equilibrium and compatibility conditions, a numerical mono-dimensional model is proposed to study the behaviour of R/C tensile members in the first cracking stage. By considering both bond behaviour and cohesion in the cracked cross-section, it is possible to enlighten their importance in the first stable or unstable cracked stage. Furthermore, the model is used to compute the minimum reinforcement ratio required for static reasons. The outcomes state that:

- Due to the bond behaviour, the minimum reinforcement ratio ρ_{min} is affected by the diameter of the reinforcing bar;

- Moreover, if the mechanical properties of concrete are constant, ρ_{min} is not affected by a structural size effect;
- On the contrary, by introducing a size effect on the material properties, ρ_{min} reduces with the size B of the tendon.

These results can not be directly extended to flexural members (i.e. by means of a model of two axially stressed elements substituting the beam). In fact, the studies on tensile and flexural members in the final cracking stage suggest that cohesion and bond have a different importance in the two cases. For these reasons the mono-dimensional model for tension is extended and enhanced for beams in bending. The results show that:

- Due to bond between steel and concrete, ρ_{min} is strongly affected by the chosen bar diameter $\varnothing$;
- The minimum reinforcement ratio ρ_{min} reduces with the height H of the cross section. On the contrary, this structural size effect does not appear if the base B of the section varies;
- A material size effect on tensile strength further reduces ρ_{min} with the height of the beam.

In conclusion, when dealing with the minimum reinforcement ratio, the proposed models are not a straightforward tool, although they could be useful to check the validity of simplified approaches.

REFERENCES

1. Comitè Euro-International du Bèton, CEB-FIP Model Code 1990 (CEB Bulletin d'Information 213-214). Thomas Telford Service Ltd., London, 1993.
2. Commission of the European Communities, Industrial Process - Building and Civil Engineering, Eurocode 2 - Design of concrete structures - Part 1, General rules and rules for buildings, Brussel, 1990.
3. Mörsch, E. (1909). *Concrete-steel construction*. Mc Graw-Hill, New York.
4. De Groot, A.K., de Kusters, G.M.A. and Monnier, Th. (1981) *Heron* **26**, 1.
5. Lutz, L.A. (1970) *ACI Journal* **67**, 778.
6. Ngo, D. and Scordelis, A.C. (1967) *ACI Journal* **64**, 152.
7. La Mendola, L. (1997) *Journal of Engineering Mechanics* (ASCE) **123**, 758.
8. Avalle, M., Ferretti, D., Iori, I. and Vallini, P. (1994). In: *Computational modelling of concrete structures*, M. Mang, H. Bicaninc and R. de Borst (Eds.). Pineridge Press, Innsbruck, pp. 723-734.
9. Yannopoulos, P.J. and Tassios, T.P. (1991) *ACI Structural Journal* **88**,3.
10. Hsu, T.T.C. (1988) *Journal of Structural Engineering* (ASCE) **85**, 2576.
11. Vecchio, F.J. and Collins, M.P. (1986) *ACI Journal* **83**, 219.
12. Bazant, Z.P. (1976) *Journal of Engineering Mechanics* (ASCE) **102**, 331.
13. Gijsbers, F. P. and Hehemann, A. A. (1977). Report BI-77-61, Institute TNO for building materials and building structures, I.B.B.C. , Delft, (in Dutch).
14. Scott, R.H., and Gill, P.A.T. (1987) *The Structural Engineer* **65**, 39.
15. Somayaji, S. and Shah, S.P. (1981) *ACI Journal* **78**, 217.
16. Press, W.H., Teukolsky, S.A., Vetterling, W.T. and Flannery, B.P. (1992). *Numerical Recipes in C: The Art of Scientific Computing*. Cambridge University Press, Cambridge, U.K.
17. Monti, G., Filippou, F.C. and Spacone, E. (1997) *Journal of Structural Engineering* (ASCE) **123**, 614.
18. Iori, I. and Rosati, G. (1994). In: *XXIII Convegno Nazionale A.I.A.S.*, Cosenza, (in Italian).
19. Maier, G. (1969). In: *Proceedings of International Union of Theoretical and Applied Mathematics Symposium on Instability of Continuous Systems*, Springer Verlag , Herrenhalb, Germany: pp. 411-417.
20. Weibull, W. (1939). *A Statistical Theory for the Strength of Materials*. Stockholm: Swedish

Royal Institute for Engineering Research.

21. Bazant, Z.P. (1984) *Journal of Engineering Mechanics* (ASCE) **110**, 518.
22. Carpinteri, A., Chiaia, B. and Ferro, G. (1995). Report 51, Department of Structural Engineering, Politecnico di Torino, Turin, Italy.
23. Carpinteri, A. and Chiaia, B. (1995) *Materials and Structures* **28**, 435.
24. Carpinteri, A. and Ferro, G. (1994) *Materials and Structures* **27**, 563.
25. Shah, S.P., Swartz, S.E. and Ouyang, C. (1995). *Fracture mechanics of concrete: application of fracture mechanics to concrete, rock, and other quasi-brittle materials.* John Wiley & Sons, Inc., New York.
26. Giuriani, E. and Rosati G. (1984) *Studi & Ricerche* **6**, 119, (in Italian).
27. Bosco, C., Carpinteri, A. and Debernardi, P.G.(1990) *Engineering Fracture Mechanics* **35**, 665.
28. Giuriani, E. and Rosati G. (1984) *Studi & Ricerche* **6**, 151, (in Italian).
29. Planas, J., Riuz, G. and Elices, M. (1995). In: *FRAMCOS-2*, F. H. Wittmann (Ed.). AEDIFICA-TO Publishers, Freiburg, pp. 1179-1188.
30. Gerstle, W.H., Dey, P.P., Prasad, N.N.V., Rahulkumar, P. and Xie, M. (1992) *ACI Structural Journal* **89**, 617.
31. Liaw, B.M., Jeang, F.L., Du, J.J., Hawkins, N.M. and Kobayashi A.S. (1990) *J. of Eng. Mech.* (ASCE), **116**, 429.
32. Bosco, C. and Carpinteri, A. (1992). In: *Applications of Fracture Mechanics to Reinforced Concrete*, A. Carpinteri (Ed.). Elsevier Applied Science, London, pp. 347-377.
33. Baluch, M.H., Azad, A.K. and Ashmawi, W. (1992). In: *Applications of Fracture Mechanics to Reinforced Concrete*, A. Carpinteri (Ed.). Elsevier Applied Science, London, pp. 413-436.
34. Bazant, Z.P. and Cedolin, L. (1980) *J. of the Eng. Mech. Div.* (ASCE) **106**, 1257.
35. Riva, P. and Plizzari, G.(1992). In: *Proceedings of Int. Conference Bond in Concrete*, CEB (Ed.). Riga, pp. 2.44-2.53.
36. Giuriani, E. and Rosati G. (1987) *Studi & Ricerche* **9**, 107, (in Italian).
37. Giuriani, E., and Ronca, P. (1979). In: *VII Convegno Nazionale A.I.A.S.*, Cagliari, pp. 6.55-6.68, (in Italian).
38. Jones, R., Swamy, R. N., Bloxham, J. and Bouderbalah, A. (1980) *The International Journal of Cement Composites* **2**, 91.
39. Plauk, G. and Hees, G. (1981). In: *IABSE Colloquium on Advanced Mechanics of Reinforced Concrete*, Delft, pp. 655-671.
40. Fantilli, A.P., Ferretti, D., Iori, I. and Vallini, P. (1998) *J. of the Structural Engineering* (ASCE) **124**, 1041.
41. Carpinteri A. (1984) *J. of the Structural Engineering* (ASCE) **110**, 544.
42. Maldague, J.C. (1965) *Annales de l'ITBTP* **213**, 1170, (in French).
43. American Concrete Institute, Building Code Requirements for Reinforced Concrete (ACI 318-95), Detroit, Michigan, 1995
44. Rokugo K., Uchida Y. and Koyanagi W. (1992). In: *FRAMCOS-1*, Z. P. Bazant (Ed.). Elsevier Publishers, London, pp. 775-781.

SIZE EFFECTS ON THE BENDING BEHAVIOUR OF REINFORCED CONCRETE BEAMS

R. BRINCKER, M.S. HENRIKSEN, F.A. CHRISTENSEN and G. HESHE,
Aalborg University, Sohngaardsholmsvej 57,
9000 Aalborg, Denmark

ABSTRACT

Load-deformation curves for reinforced concrete beams subjected to bending show size effects due to tensile failure of the concrete at early stages in the failure process and due to compression failure of the concrete when the final failure takes place. In this paper these effects are modelled using fracture mechanical concepts, and size effects of the models are studied and compared with experimental results.

KEYWORDS

Size effect, bending behaviour, reinforced concrete, minimum reinforcement, rotational capacity, fracture mechanics.

INTRODUCTION

The considered problem is the bending behaviour of simply supported concrete beams, Figure 1. The basic variables are the beam geometry given by the width b, the depth h and the span l, the concentrated load F acting at the middle of the beam and the corresponding displacement u.

The bending response of the beam is described by the load-deflection curve, Figure 1, depicting the loading force F as a function of the displacement u.

It is assumed, that the parameters influencing the bending response of the beam, apart from the basic parameters mentioned above, are the concrete type described by Young's Modulus E and the softening relations in tension and compression, the reinforcement type given by the stress-strain relation for the steel and the shear friction stress τ_f for the debonding, the reinforcement ratio φ, the number of re-bars n and the placement of the re-bars as given by the distance h_{ef} from the top of the beam. Note, that no softening relation is considered for the reinforcement. Thus, in this analysis, the contribution from necking of the re-bars, is neglected.

At early stages of the failure process, the response is governed by the tensile properties of the concrete, the elastic properties of the reinforcement steel, and the debonding process between concrete and steel. Typically, the response will show a local force maximum F_t where the concrete

A: *Test case*

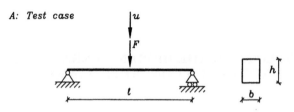

B: *Response curve*

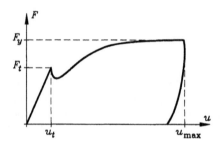

Figure 1. Fundamental problem of the investigation. A: The test case, B: Response curve.

starts cracking, a decrease afterwards, and then a slowly increasing response as the reinforcement starts debonding taking over the stresses relieved by concrete tensile failure. Later, when the tensile stresses in the concrete have decreased to zero, and yielding of the reinforcement bars is fully developed, the response curve reaches a nearly constant value F_y. For convenience, here F_y is just defined as the maximum value of the response in the "yielding regime".

Since concrete tension failure is highly size dependent, the first part of the response curve will show strong size effects. For large beams, the concrete contribution will be small and brittle compared to smaller beams, meaning that the ratio F_t/F_y will be size dependent. Thus, the ratio F_t/F_y is a central parameter for description of the size effects at early stages of the failure process. It describes, one can say, the size effect on the load scale. In most standards, the minimum reinforcement requirements aim at keeping this ratio below a certain value, securing a ductile behaviour of the beam in load control. Therefore, size effects at early stages of the failure process are closely associated with the minimum reinforcement issue.

In the first main section of this chapter, size effects on the load scale are studied under different assumptions using a non-linear fracture mechanical model for the tension failure of the concrete, and a simple friction model for the reinforcement debonding. Size effects on the bending response are studied for constant reinforcement ratios and constant values of Carpinteri's brittleness number.

When studying size effects on the late stages of the failure process, it is necessary to focus on the deformation scale. Thus, it might seem natural to choose a corresponding set of displacement parameters u_t and u_{max}, see Figure 1, and then define the corresponding ratio as the key parameter. However, since both displacements are influenced by the elastic response of the beam, in this chapter a non-dimensional parameter θ describing size effects on the displacement scale is defined by the work equation

$$\theta M_y = \int_0^\infty F \, du \tag{1}$$

where M_y is the yield moment corresponding to the yield force F_y. Now, using the simple relationship $M_y = \frac{1}{4} F_y l$, the parameter θ is given by

$$\theta = \frac{4}{l} \int_0^\infty \frac{F}{F_y} \, du \tag{2}$$

As it appears, the integral has the dimension of length, describing the total plastic deformation of the beam. This measure is not influenced by elastic contributions and is non-sensitive to the tension failure behaviour of the concrete as long as the contribution to the area under the response curve is small.

The geometric interpretation of the parameter θ is the total concentrated rotation at the yielding section under the loading force. Thus, it is a measure of the rotational capacity of the beam.

In the last main section of this chapter, size effects on the rotational capacity of concrete beams are studied using a semi-classical approach for the lightly reinforced case, where the rotational capacity is controlled by the number of cracks in the tension side of the beam, and using a fracture mechanical approach for the heavily reinforced regime, where the rotational capacity is controlled by compression failure in the concrete.

In the following model results are compared with experimental results from a large number of tests on concrete beams of different sizes, different types of concrete (normal strength, high strength), and different reinforcement ratios. The experimental results are described in Appendix A.

MINIMUM REINFORCEMENT

A commonly accepted idea behind minimum reinforcement requirements has for a long time been that the load corresponding to tension failure of the concrete F_t should be smaller than the load-bearing capacity F_y of the cracked beam section, thus

$$F_y > F_t \tag{3}$$

Now, approximating F_t by the bending strength, according to a Navier distribution of the stresses (not a good approximation) and neglecting the reinforcement contribution, yields $F_t \frac{l}{4} = f_t \frac{1}{6} b h^2$ where f_t is the tensile strength of the concrete. Using the approximate formula for the load-bearing capacity of the cracked beam section $F_y \frac{l}{4} = k h A_r f_y$, where k is a factor typically in the range $0.8 - 0.9$, and defining the reinforcement ratio as usual by $\varphi = A_r / A_c$, Equation (3) can then be written as

$$\varphi > \frac{1}{6k} \frac{f_t}{f_y} = \varphi_{\min} \tag{4}$$

This might be seen as the background for most minimum reinforcement requirements in the codes. The requirement states that, for a certain choice of steel and concrete, the reinforcement ratio should just be chosen larger than a certain value. This kind of classical reinforcement criteria are investigated in detail in the following subsection.

The weakness of the classical kind of reinforcement criteria is the simplicity of the Navier type of bending strength that does not include any size effects. Thus, using the criterion (4) in practice a value for the tension strength f_t must be used that somehow includes the size effect. Usually this is done by using a value of the tensile strength estimated from bending tests on moderate size beams. A totally different approach would be to express a criterion for stable crack growth across a reinforced concrete section. Using linear fracture mechanics and some simplifications, this leads to a criterion of the type

$$N_P > N_{\min} \tag{5}$$

where N_P is Carpinteri's brittleness number for reinforced concrete

$$N_P = \frac{f_y \sqrt{h}}{K_{Ic}} \varphi \tag{6}$$

and where K_{Ic} is the fracture toughness of the concrete. In the second subsection, the criterion given by (5) is investigated assuming that the fracture toughness can be approximated by $K_{Ic} = \sqrt{EG_F}$.

Both the classical reinforcement criterion given by Equation (4), and the fracture mechanical criterion given by Equation (5) are investigated by simulating the failure responses using a fictitious crack model for the concrete failure and a pure frictional model for the reinforcement-concrete debonding, and finding the minimum values of the reinforcement ratio φ and the brittleness number N_P corresponding to fulfilment of Equation (3).

Classical Requirements for Minimum Reinforcement

The classical requirements for minimum reinforcement are investigated by simulating the failure response of beams of different size using a fictitious crack model for the concrete tension failure and a friction model for the reinforcement-concrete debonding. Details of the investigation are given in Appendix B.

Three different concretes are considered: a normal strength concrete with linear softening, a normal strength concrete with bi-linear softening, and a high strength concrete with bi-linear softening. The softening relations are shown in Figure 5 in Appendix B.

The size of the "valley" on the response curve often seen in experimental results, occurring just after the local maximum F_t corresponding to the concrete tension failure, is totally controlled by the shear friction stress τ_f. In the limit where the shear friction stress becomes infinite, the response follows a "master curve", and cases with a finite shear friction stress appear as deviations from this curve, Figure 4, Appendix B. In the simulations, the shear friction stress was chosen as $\tau_f = 5\ MPa$ for all cases, a typical value reported in the literature for ribbed reinforcement.

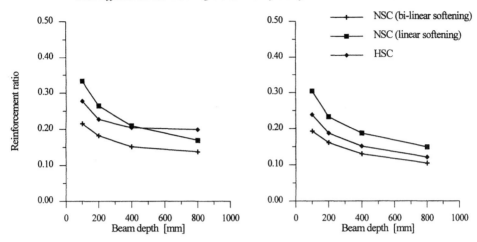

Figure 2. Reinforcement ratios corresponding to $F_y = F_t$ for different beam sizes and different softening relations. Left: No initial crack. Right: Initial crack with a depth of 7.5 % of the beam depth.

Using the simulation model so defined, the minimum reinforcement ratio φ corresponding to the smallest ratio that satisfies the overall static requirement given by Equation (3) is determined for different beam sizes. The results are shown in Figure 2. The diagrams show the results for two cases: the case of no initial crack, and the case where an initial crack is present with an initial depth of 7.5 % of the beam depth.

As it appears from the results, the reinforcement ratio corresponding to $F_y = F_t$ is clearly dependent upon the size of the beam as well as the type of softening relation. The minimum reinforcement ratio (for this type of steel) is in range $0.20 - 0.35$ for very small beams reducing to $0.10 - 0.20$ for large beams. Further, as expected, high strength concrete requires a higher minimum reinforcement than normal strength concrete, and a more ductile softening curve (the linear softening relation for the normal strength concrete has the same tensile strength and the same critical crack opening) requires a higher minimum reinforcement.

There is also an influence of the initial crack. As it appears, with an initial crack the minimum reinforcement ratio is less sensitive to the type of softening relation, and there is an overall tendency for the required minimum reinforcement ratio to become smaller. The presence of initial cracks does not seem to influence the size effect too much, although there seems to be a little less size effect in the case of initial cracks.

The overall result is that a unique minimum reinforcement ratio is difficult to define since the ratio corresponding to $F_y = F_t$ varies considerably. However, there are at least two important arguments in support of the simple criterion $\varphi > \varphi_{min}$.

First, as the results clearly indicate, if a minimum reinforcement ratio φ_{min} has been determined from bending tests on small beams, it will be on the safe side to use the criterion $\varphi > \varphi_{min}$ on

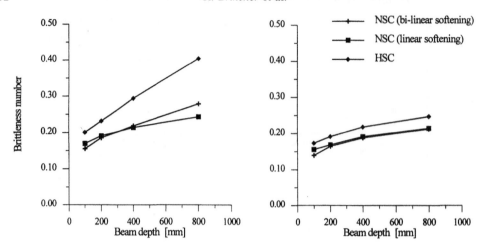

Figure 3. Carpinteri's brittleness number for reinforced beams corresponding to $F_y = F_t$ for different beam sizes and different softening relations. Left: No initial crack. Right: Initial crack with a depth of 7.5 % of the beam depth.

larger beams.

The second argument is a little more delicate, but in fact even more important. If the length scale of the softening relation is scaled with the size of the structure, i.e. if the shape of the relation is constant, and if the critical crack opening $w_c = \alpha_1 h$, where α_1 is a constant and h stands for the size of the structure, then the failure response curve (the $F - u$ curve) is shape invariant, and thus, the reinforcement ratio corresponding to $F_y = F_t$ becomes size independent. Now, a good question is, if it is reasonable to assume something like $w_c = \alpha_1 h$ in practice. It might very well be a reasonable assumption since the maximum aggregate size d_{max} typically increases with the size of the structure, and as an average it might be reasonable to assume $d_{max} = \alpha_2 h$. Further, micro mechanical considerations as well as some experimental results support the approximate relation $w_c = \alpha_3 d_{max}$. Thus, in practice taking into account the typical application of larger aggregate sizes in larger structures, the reinforcement ratio that corresponds to $F_y = F_t$ will be less size-dependent than indicated by the results of this investigation.

Carpinteri's Brittleness Number

Application of Carpinteri's brittleness number is investigated in a similar way by determining the smallest value of the brittleness number that gives a simulated response curve satisfying the overall static requirement given by Equation (3). Details are given in Appendix B. The results of the investigation are shown in Figure 3.

The overall impression is that there is a size effect on Carpinteri's brittleness number. The brittleness number increases with the beam size. This is to be expected, since the square root size dependence is the strongest possible dependence corresponding to the linear fracture mechanical case. The tensile bending problem is known to be influenced by fracture mechanical non-linearities leading to a weaker influence of the size than predicted by the brittleness number. Thus, when the minimum reinforcement ratio is corrected by a square root factor, it is to be expected that this will over-correct the size effect on the minimum reinforcement ratio, and therefore, the brittleness number becomes size depending itself.

As it appears, there is a large difference between the case of no initial crack, and the case of an initial crack. In the case with no initial crack, there is a relatively strong size effect on the brittleness number, whereas this size effect is significantly smaller in the case of an initial crack. In the case of an initial crack, the size effect on the brittleness number is smaller than the size effect on the reinforcement ratio. Thus, since initial cracks must be assumed to be present in real structures, it seems that the brittleness number for this reason might be a better choice for minimum reinforcement requirements than the reinforcement ratio.

On the other hand, a critical brittleness number found from tests on small beams will lead to results for minimum reinforcement on large beams that are on the unsafe side.

Further, it seems like the brittleness number is less sensitive to the shape of the softening relation, but depends more on the strength level of the concrete.

ROTATIONAL CAPACITY

In the lightly reinforced regime, the rotational capacity is controlled by the number of cracks and the local debonding and yielding of the reinforcement around each crack. If no debonding takes place (case of infinite shear friction stress τ_f), the length over which yielding takes place tends to zero, and thus, the contribution from yielding of the reinforcement tends to zero. In this case however, the number of cracks becomes large and thus, so does the contribution to the total work from the cracking of the concrete. In the extreme case of a very small friction stress, only one crack develops, and thus, yielding of the reinforcement is mainly responsible for maintaining the rotational capacity. These effects are described in the next subsection.

In the heavily reinforced regime, the rotational capacity is controlled by the compression failure of the concrete. A large part of the rotational capacity might come from plastic deformations of the reinforcement steel, but the deformation capacity of the concrete is the limiting parameter. If the concrete has a small deformation capacity the contribution from the plastic deformation of the reinforcement as well as the contribution from the concrete itself becomes small. The rotational capacity of heavily reinforcement beams is studied in the second subsection using a fracture mechanical approach for the compression failure of the concrete.

Lightly Reinforced Beams

In this case a semi-fracture mechanical approach is used. The location of cracks and the number of cracks are determined by a classical approach, i.e. the concrete cracks when the tensile strength is reached, and the response (the $F - u$ curve) is calculated assuming a constant shear friction stress over the debonded area. Once the response is determined, the rotational capacity is determined as described by Equation (2). A semi-fracture mechanical approach is introduced by adding the energy dissipated in the tensile cracks to the total work determined as the integral of the simulated response (classical contribution). The energy dissipated in the tensile cracks is estimated as nA_cG_F where n is the number of cracks. Thus, the model assumes the cracks to be fully opened and the cracks to extend approximately over the whole beam area A_c. Details are given in Appendix C.

Using a model like this, the rotational capacity is strongly dependent on the strain hardening of the reinforcement. If the strain hardening is small, yielding takes place only over a small length of the re-bars around each crack, and the only way of extending the yield length of the re-bars, is to increase the strain hardening of the steel, i.e. the ratio f_u/f_y where f_u is the ultimate failure stress of the reinforcement. This effect is shown in Figure 3 in Appendix C.

Another important main result is that the model does not show a so strong dependency on the shear friction stress τ_f as one would expect, Figure 4 in Appendix C. This is due to the semi-fracture mechanical approach where the increasing values of τ_f reduce the contribution from yielding of the reinforcement, but at the same time increase the contribution from the energy dissipation in the tensile cracks.

Figure 4 shows the size effects for two different values of the shear friction stress τ_f. If the shear friction stress has a small or moderate value (left part of Figure 4) then the number of cracks is moderate too, and the influence from the tensile failure on the rotational capacity is small. In this case, the rotational capacity is dominated by the classical contributions from yielding and frictional debonding which does not show size effects, and thus, the size effect is small. However, if the shear friction stress becomes large (right part of Figure 4), the number of cracks increases, and so does the contribution to the rotational capacity from dissipation of energy in the tensile cracks. Thus, since this contribution is size dependent, the total rotational capacity becomes size dependent. As it appears from the results, the shear friction stress has to be very large in order to enforce a size effect of importance, and even in that case, the size effects are still moderate.

Figure 6 in Appendix C shows the rotational capacity for the two different ribbed reinforcement types used in the experimental investigation, a high deformation capacity type, and a low deformation capacity type. In this case the shear friction stress is taken as $\tau_f = 5\ MPa$. As expected, size effects are small for both types of reinforcement, and the deformation capacity of the reinforcement has a large influence on the rotational capacity of all beams.

Heavily Reinforced Beams

For this case a fracture mechanical description of the compression failure is used. The main argument for this is that experimental results show that the post-peak behaviour including the post-peak energy dissipation is relatively independent of the length of a compression specimen, see Figure 2 in Appendix D.

Naturally, this leads to an assumption of localized compression failure, thus, compression failure is assumed to take place over a certain length of the compression zone. This length is denoted the characteristic length l_{ch}. The characteristic length is assumed to be proportional to the depth of the compression zone corresponding to a compression-shear failure mode with the development of slip-planes at a certain angle to horizontal. In this case the characteristic length becomes proportional to the depth h_c of the compression zone $l_{ch} = \beta h_c$. Assuming an inclination angle of the slip-planes similar to the cone-failure of a cylinder in compression, justifies the expectation of β having a value around 4.

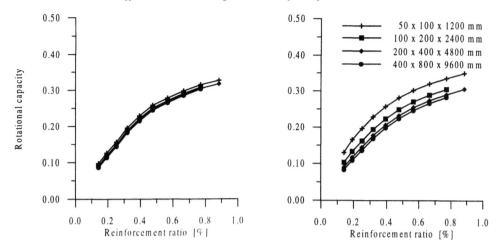

Figure 4. Size effects on the rotational capacity in the lightly reinforced regime for different values of the shear friction stress τ_f. Left: Shear friction stress $\tau_f = 5\ MPa$. Right: Shear friction stress $\tau_f = 20\ MPa$.

The model assumes a simple linear softening relation, so the model has two main parameters: the inclination parameter β, and the critical deformation w_c of the concrete in the compression failure zone.

The model is roughly calibrated by simulating failure responses of beams, calculating the rotational capacity according to Equation (2) and comparing with the experimental results for normal strength concrete. It turns out that in order to have reasonable agreement with the experimental results, the values of the key parameters of the model have to be chosen as $w_c \cong 4$ mm, $\beta \cong 8$. These values are somewhat higher than expected, but the rather high values might be interpreted as a consequence of the loading arrangement for the tests where the loading plate used for distribution of the concentrated load might act as confinement.

In Appendix D a simple model for tension failure of the reinforcement was included in order to check if the model showed a reasonable switch between concrete compression failure and reinforcement tension failure. Results for pure compression failure are shown in Figure 5.

As it appears, the model shows a clear size effect on the rotational capacity, the rotational capacity becoming significantly smaller with increasing beam size. Further, for all beam sizes, the rotational capacity predicted by the model decreases with the reinforcement ratio. The influence from both the size and the reinforcement ratio is rather large. Thus, for high values of the reinforcement ratio, the rotational capacity is small for all beam sizes, and for the large beams, the rotational capacity is small also for moderate values of the reinforcement ratio.

The results for the high strength concrete are obtained by using the same values of w_c and β as for

R. Brincker et al.

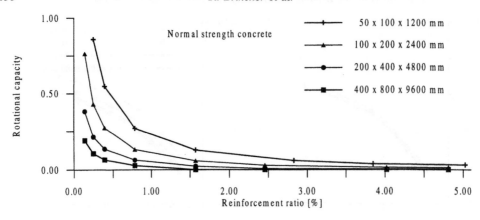

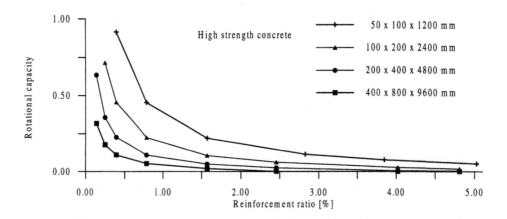

Figure 5. Rotational capacity in the heavily reinforced regime. Top: Normal strength concrete. Bottom: High strength concrete.

Rotational capacity

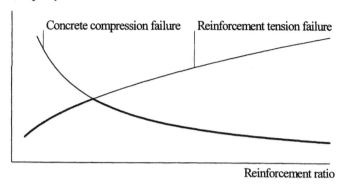

Figure 6. Combining models for lightly reinforced and heavily reinforced beams.

the normal strength concrete; only the compression strength and the failure strain (peak strain) have been increased. For higher strength concrete, there is no guarantee, that the key parameters w_c, β will be the same. In fact, experimental results (see Appendix D) indicate that a smaller value of w_c has to be used. However, keeping the key parameters the same, the results show a clear increase of the rotational capacity with strength.

The rotational capacity described by the model also depends weakly on the yield strength of the reinforcement just like the results of the model for the lightly reinforced regime described in the preceding subsection depends weakly on the tensile strength of the concrete. However, considering the two models defined above, the model for the lightly reinforced regime is describing rotational capacity mainly controlled by reinforcement tension failure, and the model for the heavily rein-forced regime is describing rotational capacity mainly controlled by concrete compression failure. Now, Taking the smallest value of the rotational capacity predicted by the two models defines a model valid for all reinforcement ratios, the intersection point defining the transition from the failure mode where the rotational capacity is controlled mainly by reinforcement tensile properties to the failure mode where the rotanional capacity is controlled mainly by concrete compression properties, see Figure 6.

APPENDIX A:

TEST PROGRAMME FOR BENDING FAILURE OF REINFORCED CONCRETE BEAMS OF DIFFERENT SCALE

Prepared by M.S. Henriksen, R. Brincker and G. Heshe

Abstract

In this appendix a brief summary of experiments on reinforced concrete beams in three-point bending performed at Aalborg University is given. The aim of the investigation is to determine the full load-deflection curves for different beam sizes, different types of concrete and different amounts and types of reinforcement, and to calculate the rotational capacity for all beams. The rotational capacity is here defined as the non-dimensional area under the load-deflection curve, and the results are shown as functions of the reinforcement ratio.

Introduction

This appendix gives a summary of the experimental results from tests of reinforced concrete beams in three-point bending carried out in the Structural Research Laboratory at the Department of Building Technology and Structural Engineering, Aalborg University. The experiments were performed in the period of August 1994 to June 1995 in connection with a Round Robin on "Scale Effects and Transitional Failure Phenomena of Reinforced Concrete Beams in Flexure" initiated by the European Structural Integrity Society - Technical Committee 9. The test programme has been designed according to the proposals given by Bosco and Carpinteri [1]. The aim of the investigations was to verify the scale dependency of plastic rotational capacity and minimum reinforcement and the existence of transitional phenomena of failure.

The test programme at Aalborg University has involved 114 reinforced concrete beams for determination of load-deflection curves, 54 plain concrete beams for determination of the fracture energy G_F and 324 concrete cylinders for determination of strength parameters. For the reinforced concrete beams four different parameters were varied. The slenderness ratios were 6, 12 and 18 and the beam depth's were 100 mm, 200 mm and 400 mm giving a total of nine different geometries. In order to fulfill the requirements for the Round Robin 6 reinforcement ratios were chosen between 0.06 % and 1.57 % giving a lightly and more heavily reinforced regime. For the concrete both a normal strength and a high strength concrete were chosen with a compressive strength of approximately 60 MPa and 100 MPa. All experiments were repeated three times. In Henriksen et al. [2] a more detailed description of the experiments is given.

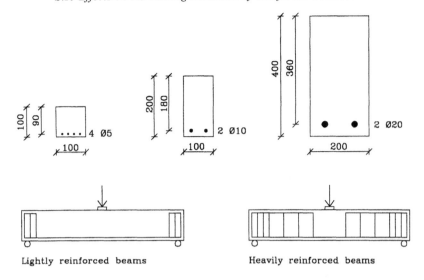

Figure A1. Scaling of cross-section and design of a typical test beam. Units in *mm*.

Test Specimens

The geometry of the beams was chosen in accordance with the requirements given by Bosco and Carpinteri [1], and the different geometries and main reinforcements for all beams are given in Table A1. The reinforcement ratios (the steel area divided by cross-section of the beam) were varying from 0.06 % to 1.57 % for the normal strength concrete beams and 0.14 % to 0.39 % for the high strength concrete beams.

The reinforcement of the beams was designed so that all beams failed in bending. Some of the highly reinforced beams were reinforced with stirrups to avoid anchorage and shear failure. Furthermore, to avoid influence on the compression failure, stirrups and compressive reinforcement were not placed in the mid zone of the beams.

The geometry of the cross-sections of the beams is shown in Figure A1. The distance from the top of the beam to the middle of the reinforcement bars were in all cases equal to $0.9h$, thus $h_{ef} = 0.9h$ for all beams. Note that the beams and the size of the reinforcement bars are scaled to preserve geometrical similitude.

Concrete

Two types of concrete were used for the experiments. The mix of concrete was prepared in accordance with the requirements given by Bosco and Carpinteri [1] and the recipes are listed in Table A2. Due to the large amount of concrete for each casting, a commercial manufacturer (ISO 9002 certificate) delivered the concrete. Totally 18 castings have been carried out. The test beams were cast in steel moulds.

Table A1: Overview of test beams, units are in [mm]

$b \times h$	l/h	NSC	NSC and HSC			NSC	
		0.06 %	0.14 %	0.25 %	0.39 %	0.78 %	1.57 %
100 × 100	6					4 ø5	8 ø5
100 × 100	12		1 ø4	2 ø4	2 ø5	4 ø5	8 ø5
100 × 100	18					4 ø5	8 ø5
100 × 200	6					2 ø10	4 ø10
100 × 200	12		1 ø6	1 ø8	1 ø10	2 ø10	4 ø10
100 × 200	18					2 ø10	4 ø10
200 × 400	6					2 ø20	4 ø20
200 × 400	12	1 ø8	1 ø12	1 ø16	1 ø20	2 ø20	4 ø20
200 × 400	18					2 ø20	4 ø20

Table A2: Mix proportions of the two types of concrete

Content	Normal strength concrete		High strength concrete	
Units	kg/m^3	l/m^3	kg/m^3	l/m^3
Cement	350	111	466	148
Water	160	160	146	146
Silica	0	0	36.1	16.4
Plasticiser 1	3.84	3.2	1.80	1.5
Plasticiser 2	0	0	12.6	10.4
Sand (0-2 mm)	898	335	899	324
Gravel (4-8 mm)	896	335	899	324
Air	0	50	0	0
Density	2307 kg/m^3		2399 kg/m^3	

The mechanical properties of the concrete were obtained by standard tests. The tensile splitting strength, the compressive strength, the compression softening and the modulus of elasticity were determined on cylinders (diameter 100 mm and depth 200 mm). The bending tensile strength and the bending fracture energy were determined from tests on notched RILEM beams with a span of 800 mm, a thickness of 100 mm and a depth of 100 mm. The material parameters are listed in Table A3.

For both normal strength and high strength concrete experiments on cylinders (100 × 200 mm) to obtain the stress-strain relations in uniaxial compression have been carried out by Dr. Henrik Stang at the Technical University of Denmark. Typical compression softening curves are shown in Figure A2. Note the more brittle behaviour of the high strength concrete.

Reinforcement

In order to have six different reinforcement ratios and to reinforce beams of different scale without changing the geometry of the cross-sections it was necessary to use reinforcement with 8 different diameters. Two types of ribbed reinforcement bars were used with approximately the same type of ribs. The ø4 mm and ø5 mm steel bars were cold deformed and had a relatively small deformation

Table A3: Mechanical properties of the two types of concrete

	Units	Normal strength concrete		High strength concrete	
		Mean	S. Dev.	Mean	S. Dev.
Compressive strength	[MPa]	64.0	6.12	98.5	6.60
Splitting strength	[MPa]	4.09	0.54	6.06	0.28
Modulus of elasticity	[MPa]	4.23E4	0.31E4	4.55E4	0.23E4
Bending tensile strength	[MPa]	5.51	0.34	7.16	0.29
Bending fracture energy	[J/m^2]	126	8.30	118	5.24

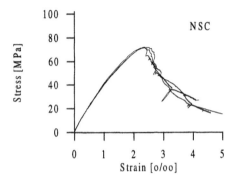 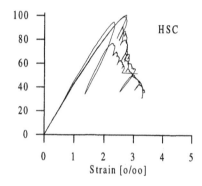

Figure A2. Compression tests with softening for normal strength concrete (left) and high strength concrete (right). The test specimen is a 100×200 ($d \times h$) mm cylinder.

capacity, while the ø6 mm, ø8 mm, ø10 mm, ø12 mm, ø16 mm and ø20 mm steel bars had a large yield capacity and a clear strain hardening.

The mechanical properties of the steel were determined on 500 mm long specimens subjected to uniaxial tension. The results given as nominal values are summarized in Table A4. It is observed from Table A4 that the bars with a large yield capacity fulfil the Eurocode 2 requirements for high ductility reinforcement $f_u/f_y < 1.08$ and $\epsilon_{su} > 5\%$. Some typical curves from the tensile test of the ribbed steel bars are shown in Figure A3. Note the difference in the behaviour for the two types of steel.

Test Set-up

All beams were subjected to three-point bending. The tests were carried out using the same specially designed servo-hydraulic testing system allowing for testing of beams of different size. At both supports horizontal displacements and rotations were allowed for and, at one support, also rotations around the beam axis were allowed. At the load point rotations were allowed around all axes. The load was transferred through a square steel plate with sides equal to the beam width.

The stroke (the displacement of the piston of the hydraulic actuator) was measured using the

Table A4: Mechanical properties of the reinforcement bars.

Steel type	Young's Modulus E_s [MPa]	Yield strength f_y [MPa]	Yield capacity $\Delta\epsilon_y$ [%]	Ultimate strength f_u [MPa]	Ultimate strain ϵ_{su} [%]	Ultimate to yield strength ratio f_u/f_y [−]
ø4	2.01E5	740	0	740	1.41	-
ø5 charge 1	1.94E5	701	0	701	2.20	-
ø5 charge 2	2.01E5	708	0	708	2.94	-
ø6	2.09E5	600	3.44	664	13.1	1.11
ø8	2.05E5	604	2.29	685	10.7	1.13
ø10	2.06E5	611	2.27	681	11.2	1.11
ø12	2.01E5	555	2.56	642	11.5	1.16
ø16	1.96E5	531	1.60	630	11.0	1.19
ø20	1.82E5	531	1.13	624	9.27	1.18

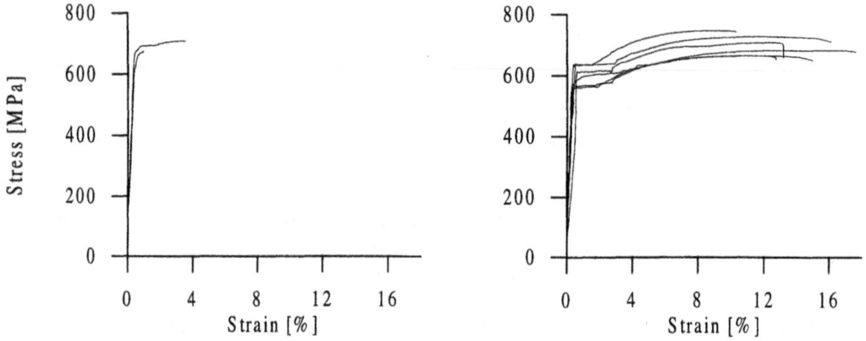

Figure A3. Typical stress-strain relations for ribbed steel bars subjected to tensile testing. Typical cold deformed bars (left) and typical normal ribbed bars (right).

built-in LVDT (Linear Variable Displacement Transformer). The vertical deflection of the beams was measured at eight points along the beam axis. The rotations of the ends of the beams were measured using two LVDT's at each end.

The mutual rotations of the cross-sections at different points along the beam axis were measured using a number of specially designed measuring frames. Dividing these rotations by the distance between the measuring frames gives a direct estimate of the average curvature between the frames.

Main Test Results

The main results for the normal strength and high strength concrete beams with slenderness ratio 12 are shown in Figures A4-A6. The results are given as load-deflection curves for the mid point of the beam.

Typical distributions of vertical displacement and of the curvature along the beam axis at different load levels (indicated at the load deflection curve) are shown in Figure A7 for a normal strength concrete beam with dimensions $200 \times 400 \times 4800 \ mm \ (b \times h \times l)$ reinforced with one ⌀20 mm ribbed steel bar corresponding to a reinforcement ratio of 0.39 %.

Rotational capacity

For each beam the rotational capacity was obtained by calculating the total plastic work as the area under the load-deflection curve and dividing this work by the maximum yield moment (the maximum moment at the yielding regime for each load-deflection curve). This value is a direct estimate of the total plastic rotation of the beam ends.

Results for the rotational capacity as a function of the reinforcement ratio are given in Figure A8 for normal strength concrete and in Figure A9 for high strength concrete.

The results for slenderness ratios 6 and 18 in Figure A8 are too few to indicate any clear size effect. In fact, the small beams have the smallest rotational capacity. This is believed to be due to the low deformation capacity of the reinforcement used for the smallest beams. The results for slenderness ratio 12, however, show a clear size effect for the two larger beam sizes. The smallest beam has a significantly lower rotational capacity, again indicating the large influence from the limited deformation capacity of the cold deformed reinforcement.

The results for high strength concrete in Figure A9 are too limited to indicate any clear size effect. Again the limited deformation capacity of the cold deformed reinforcement causes the small beams to have a significantly lower rotational capacity.

References

1. Bosco, C. and A. Carpinteri (1993): Proposals for a Round Robin on Scale Effects and Transitional Failure Phenomena of Reinforced Concrete Beams in Flexure. European Structural Integrity Society, Technical Committee 9 on Concrete, ESIS-TC9.
2. Henriksen, M.S., J.P. Ulfkjaer, and R. Brincker (1996): Scale Effects and Transitional Failure Phenomena of reinforced Concrete Beams in Flexure. Part 1. Fracture & Dynamics Paper No. 81. Department of Building Technology and Structural Engineering, Aalborg University, Denmark.

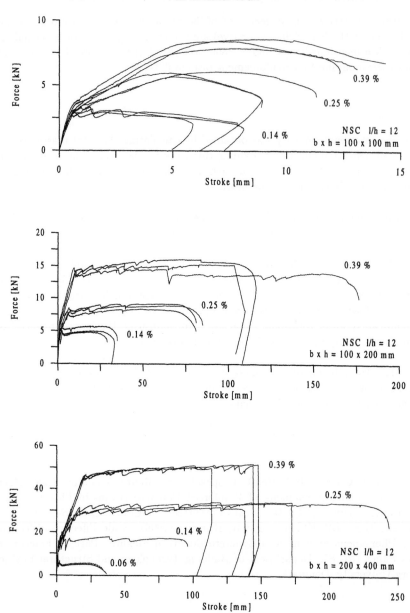

Figure A4. Load-displacements curves for normal strength concrete (NSC) beams with reinforcement ratios 0.06 %, 0.14 %, 0.25 % and 0.39 % and slenderness number 12.

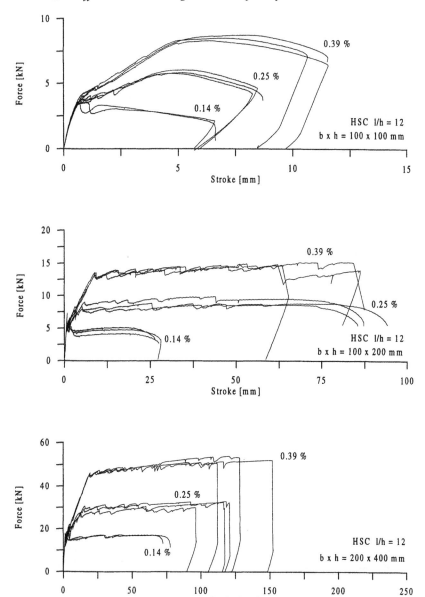

Figure A5. Load-displacements curves for high strength concrete (HSC) beams with reinforcement ratios 0.14 % 0.25 % and 0.39 % and slenderness number 12.

R. Brincker et al.

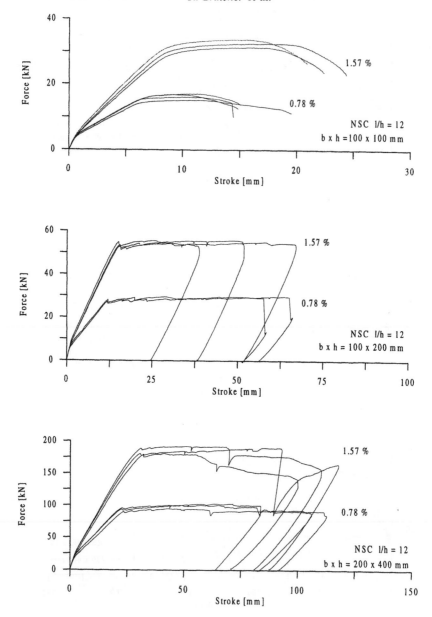

Figure A6. Load-displacements curves for normal strength concrete (NSC) beams with reinforcement ratios 0.78 % and 1.57 % and slenderness number 12.

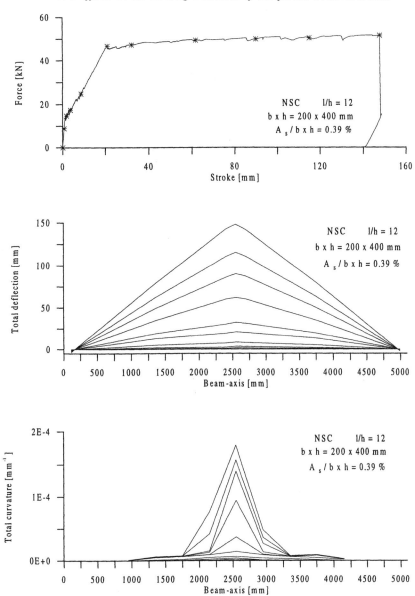

Figure A7. Load-deflection curve (top), vertical distribution (mid) and distribution of curvature (bottom) along the beam axis for a normal strength concrete beam ($200 \times 400 \times 4800\ mm$) with a reinforcement ratio of 0.39 %. The load levels are marked with asterisk (top).

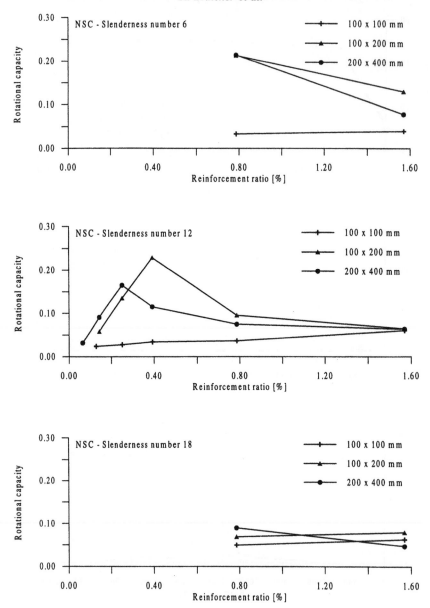

Figure A8. Rotational capacity as a function of reinforcement ratio for normal strength concrete (NSC) beams with slenderness numbers 6, 12 and 18.

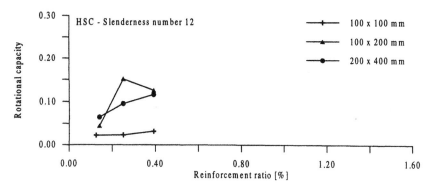

Figure A9. Rotational capacity as a function of reinforcement ratio for high strength concrete (HSC) beams with slenderness number 12.

APPENDIX B:

MODEL FOR SINGLE CRACK EXTENSION IN LIGHTLY REINFORCED BEAMS

Prepared by F.A. Christensen and R. Brincker

Abstract

In this appendix the failure behaviour of lightly reinforced concrete beams is investigated. A numerical model based on the fictitious crack approach according to Hillerborg [1] is established in order to estimate the load-deflection curve for lightly reinforced concrete beams. The debonding between concrete and reinforcement is taken into account by introducing a debonded zone with constant shear friction stress. Results are presented for material models representing normal strength concrete (two degrees of brittleness) and high strength concrete. The properties of the model are investigated and the results of the model are compared with results from experiments.

Introduction

The main purpose of this work is to formulate a model well suited for estimation of accurate load-deflection curves for lightly reinforced concrete beams. The aim is to be able to investigate which parameters are important for the stability of the tensile failure in load control. If the yield load F_y of the beam is larger than the local peak load F_t corresponding to tensile failure of the concrete, then the tensile failure is normally accepted to be considered stable. Thus, the determination of the local peak load corresponding to concrete tensile failure is essential in the analysis. It is well known, that those kinds of failure problems are governed by fracture mechanical effects, and thus, a fracture mechanical approach must be used in order to establish accurate estimates of the ratio F_t/F_y for different sizes of the structure.

Bosco and Carpinteri [2] formulated a model based on linear elastic fracture mechanics. They showed that, according to this model, the failure mode changes when the beam depth is varied, the reinforcement ratio remaining the same. Only when the reinforcement ratio is inversely proportional to the square root of the beam depth, the mechanical behaviour is reproduced by the model. Baluch et al. [3] have shown how strain softening of the concrete can be taken into account in the model proposed by Bosco and Carpinteri. Due to the limitations of linear elastic fracture mechanics (does not describe the initial stages of cracking in a reinforced structure, and does not describe the weaker size effects introduced by non-linear fracture mechanical models), several other researchers have used an approach where the main crack at the midpoint of the beam is modeled as a fictitious crack [4-6].

In this appendix, a non-linear fracture mechanical approach is adopted for the concrete tensile failure using the fictitious crack model introduced by Hillerborg [1]. The softening relations are assumed to be linear or bi-linear, and different softening parameters are used to represent normal

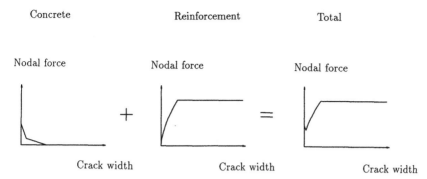

Figure B1. The relation between the nodal force and the crack width for the composite node representing both concrete and the force from the reinforcement.

strength concrete and high strength concrete. When a crack in a reinforced concrete beam starts opening, the reinforcement starts getting loaded taking over the stresses relieved by the tensile failure. For this reason, the modeling of the debonding process between concrete and reinforcement is of great importance. In this work, a simple constant shear friction model is used for the debonding stress.

Model Formulation

The model used in this investigation is based on the sub-structure method introduced by Petersson [7], and reformulated for a beam in three-point bending by Brincker and Dahl [8] in a way that makes it possible to obtain the entire load-deflection curve. Only the basic ideas of the model will be presented here. A beam subjected to three-point bending is considered. It is assumed that a crack starts to extend at the midpoint of the beam and a fracture zone develops in front of the crack tip. According to the fictitious crack model a point on the crack extension path can be in one of three possible states: 1) elastic state 2) fracture state (material is softened by microcracking) and 3) a state of no stress transmission (crack fully developed).

The method is implemented by dividing the midpoint section of the beam into a number of nodes. For each node a relation between the nodal force and the crack width is used as input to a numerical solving scheme. In order to model a reinforced concrete beam, the node located at the place of the reinforcement has to represent both the reinforcement and the concrete. Since the bond-slip between the reinforcing bars and the concrete substantially influences the response of the beam, it is necessary to take this effect into account. This is done by assuming a constant shear stress at the debonded interface between the rebars and the concrete. To simplify the problem the force of the reinforcement bar is assumed to act directly on the faces of the concrete tensile crack. The relation between nodal force and the crack width is found by superposition of the contributions from the concrete and from the reinforcement as shown in Figure B1.

Using a model like this, the complete relation between the load and the deflection at the midpoint of the beam can be estimated. An example of a load-deflection curve is shown in Figure B2 where a bi-linear softening relation has been used to represent the concrete tension failure. Stress distri-

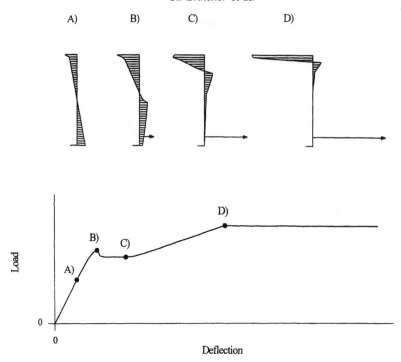

Figure B2. Bottom: The load-deflection curve divided into different parts according to the stress distribution. Top: Stress distributions, A) The tensile stress is reached at the bottom of the beam, B) peak load at first crack, C) The critical crack width is reached at the bottom of the beam, D) The yield stress is reached in the reinforcement.

butions corresponding to certain points on the load-deflection curve are also shown in Figure B2. Until point A) the beam behaviour is purely elastic and point A) represents the point where the tensile stress is reached at the tensile side of the beam. The crack starts to extend, and while the crack width is small, a zone of approximately constant failure stress will be present in the tensile side of the beam. This plastic-like stress distribution causes the load to increase until the peak load at first cracking is reached, point B). At point C) the critical crack width is reached at the bottom of the beam. When the reinforcement contribution becomes higher as the crack starts to open, the load starts to increase in the load-deflection diagram. Hereafter the load increases rapidly until the yield stress is reached in the reinforcement, point D).

The drop in load between the peak load at first cracking and the yield plateau, is characteristic for the failure response of lightly reinforced beams. At any stage of the failure process, the load carrying capacity of a lightly reinforced concrete beam can be divided into contributions from the concrete and the reinforcement respectively. For the model, these two contributions together with the total load capacity are shown in Figure B3. At small deflections the load is carried by the concrete, while at larger deflections (when the concrete is cracked) the load is carried by the reinforcement.

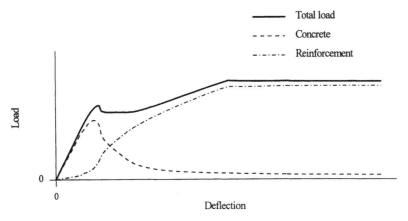

Figure B3. The load deflection curve divided into contributions from reinforcement and concrete, respectively.

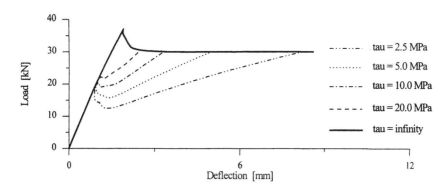

Figure B4. Load-deflection curves for the $200 \times 400 \times 4800$ mm beam of normal strength concrete. The reinforcement ratio is 0.25%. The friction stresses τ are varied from 2.5 MPa to infinity.

For a given softening relation for the concrete, the pull-out of the reinforcement (the debonding between concrete and reinforcement) strongly controls the failure behaviour of the beam after the peak load. In the model a constant shear friction stress τ_f is assumed to act on the debonded interface between concrete and reinforcement. In Figure B4, load-deflection curves are shown for a $200 \times 400 \times 2400$ mm beam where the friction stresses are varied from 2.5 MPa to infinity. As it appears, the value of the friction stress significantly influences the behaviour of the beam between the peak load at first cracking and the yield plateau. In the literature values of the friction stress are typically reported in the range 3-8 MPa. for instance, Planas [6], found about 5 MPa for ribbed reinforcement. In this interval the peak load is not influenced much, whereas the drop in load after the peak is more sensitive to changes in the value of τ_f. Note that the limit $\tau_f \rightarrow \infty$ defines a "master curve" in the sense that all failure responses for finite values of τ_f might be considered as deviations from this curve around the peak point of the master curve.

Table B1: Softening properties of normal strength and high strength concrete

Material property	Normal strength concrete (bi-linear softening)	Normal strength concrete (linear softening)	High strength concrete
Tensile strength f_t [MPa]	3.0	3.0	5.0
Youngs modulus E [MPa]	40,000	40,000	40,000
Fracture energy G_F [N/mm]	0.120	0.240	0.120

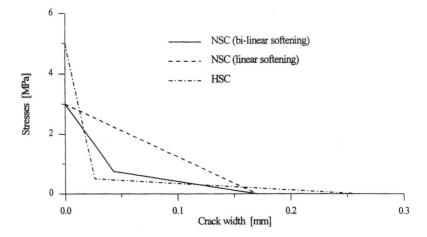

Figure B5. Softening relations for the three different materials used in the modelling.

Properties of the Model

In the following section some typical results from the model are given for a normal strength concrete with a linear and a bi-linear softening relation and a high strength concrete with a bi-linear softening relation. Material parameters representing these materials are given in Table 1 and the softening relations are shown in Figure B5. The softening parameters for the bi-linear softening relations are representative for the two types of concrete used in the experimental investigation (Appendix A). Failure responses are shown for four beam depths of 100, 200, 400 and 800 mm. The b/h-ratio is equal to 0.5 and the l/h-ratio is equal to 12. For all the simulations the debonding friction stress is 5.0 MPa and the yield stress of the reinforcement is 500 MPa.

In the two top plots in Figure B6 load-deflection curves are shown for different beam sizes of normal strength concrete with a bi-linear softening relation and a linear softening, respectively. In both cases the reinforcement ratio is equal to 0.25%. Note the strong size effect on the results and the more ductile behaviour of the beam with the linear softening relation (doubled fracture energy). The bottom plot in Figure B6 shows similar results for high strength concrete. Note the more brittle behaviour of these beams. For all three cases, keeping the reinforcement ratio constant, the ratio F_t/F_y is increasing when the beam size is decreased. This indicates, that in order to reproduce the same F_t/F_y ratio when the beam depth is increased, the reinforcement ratio must be decreased. More results from this kind of investigation are shown in the main part of the chapter.

By comparing the load-deflection curves for the normal strength concrete with bi-linear and linear softening (doubling the value of G_F) it can be observed that the reinforcement ratio should be increased when the fracture energy is increased to maintain the same F_t/F_y ratio. Further, when the tensile strength is increased the reinforcement ratio should also be increased to maintain the F_t/F_y ratio.

Some of these effects are taken into account by the brittleness number for reinforced concrete proposed by Bosco and Carpinteri [2]

$$N_P = \frac{f_y\sqrt{h}}{K_{Ic}}\varphi \tag{1}$$

where K_{Ic} is the fracture toughness of the concrete. It is assumed that the fracture toughness can be approximated by $K_{Ic} = \sqrt{EG_F}$. Varying the reinforcement ratio, the failure response can now be estimated for different values of the brittleness number N_P.

In Figure B7 the value of N_P is varied ranging from 0.15 to 0.30 for the three materials (for the $200 \times 400 \times 2400$ *mm* beam). The results show that the peak load at the first cracking F_t is equal to the yield load F_y for values of the brittleness number N_P at about 0.20 for both types of normal strength concrete. However, for the high strength concrete, the corresponding value is about 0.30. This indicates, that the brittleness number does not account for an isolated increase of the strength. This is to be expected since the brittleness number is a linear fracture mechanical parameter that does not include any description of the shape of the softening curve.

Further, It should be noticed that, although the value of the brittleness number is kept constant, the load-deflection curves for the high strength concrete beams show a much more brittle behaviour than the load-deflection curves for the normal strength beams, bottom plot, Figure B7. Again, this is due to the fact that the brittleness number does not include description of the shape of the softening relation.

Figure B8 shows how the failure response changes when the brittleness number is kept constant and the beam depth is varied. For all the three different materials it can be stated that the failure behaviour changes significantly with the beam depth. These conclusions are believed to be generally valid for the case of no initial crack. Using a non-linear approach like the fictitious crack model, the case of no initial crack can be analysed as it has been shown here. However, using linear fracture mechanics, an initial crack must be present, and thus, the basic assumptions for using the brittleness number are in principle not satisfied. A similar investigation can be carried out assuming the presence of an initial crack, and for this case, the ratio F_t/F_y becomes more stable in the case of constant brittleness number. The main results of this investigation are shown in the main part of the chapter.

Results shown in Figures B6-B8 are normalized by the yield force on the load scale and by the deformation corresponding to point A) in Figure B2 on the deformation scale.

R. Brincker et al.

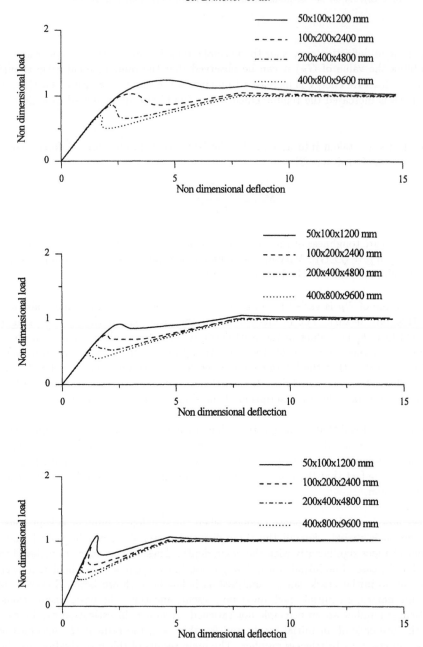

Figure B6. Size effects on the failure response for constant reinforcement ratio $\varphi = 0.25\%$. Top: Normal strength concrete with linear softening. Middle: Normal strength concrete with bi-linear softening. Bottom: High strength concrete with bi-linear softening.

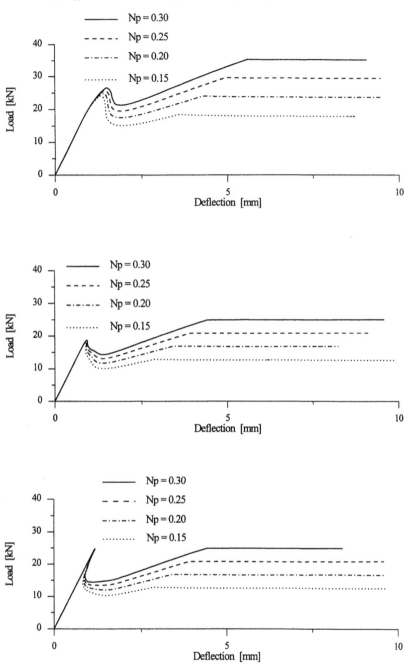

Figure B7. Variation of the failure response with brittleness number N_P for the 200 × 400 × 4800 *mm* beam. Top: Normal strength concrete with linear softening. Middle: Normal strength concrete with bi-linear softening. Bottom: High strength concrete with bi-linear softening.

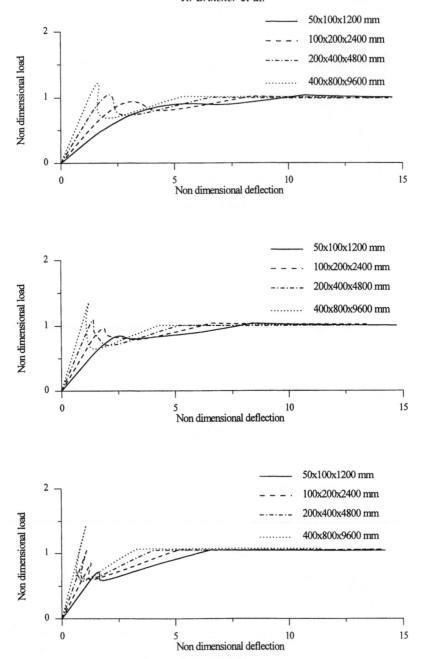

Figure B8. Size effects on the failure response for constant Brittleness number N_P. Top: $N_P = 0.20$ and normal strength concrete with linear softening. Middle: $N_P = 0.20$ and normal strength concrete with bi-linear softening. Bottom: $N_P = 0.30$ and high strength concrete with bi-linear softening.

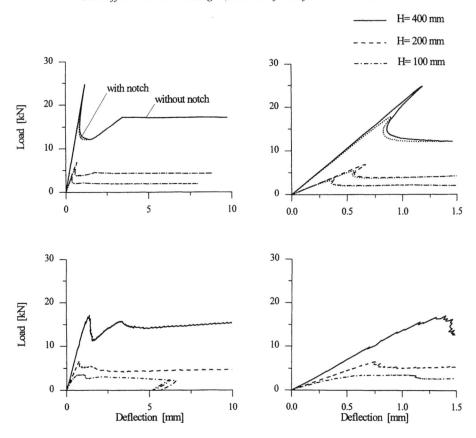

Figure B9. Failure response comparison between model and experimental results for the reinforcement ratio of 0.14 % and high strength beams. Top: Model results. Bottom: Experimental results.

Comparison with Experimental Results.

In Figure B9 model results and typical results obtained by experiments are compared for the high strength concrete and a reinforcement ratio of 0.14 %. The failure response is simulated for the case of no initial crack and for the case of an initial crack with an initial depth of 7.5 % of the beam depth.

Evaluating the model it should be noticed, that no tuning of the model has been performed. All material parameters in the model have either been taken from the experimental investigation described in Appendix A or as typical values reported in the literature.

As it appears, the model clearly overestimates the peak value in the case of no initial crack, whereas the simulated failure response shows a better agreement with experimental results if an initial crack is present. Similar conclusions can be drawn for the normal strength concrete and for other reinforcement ratios. The results indicate that an initial crack should be assumed in all experimental cases of a size of about the chosen value or may be slightly larger. The "valley" after the concrete tension failure peak seems to be well estimated by the model, indicating that it seems reasonable to assume a constant frictional shear stress in the debonded zones at each side of the crack.

References

1.   Hillerborg A., M. Modeer and P.E. Peterson (1976): Analysis of Crack Formation and Crack Growth in Concrete by means of Fracture Mechanics and Finite Elements. *Cement and Concrete Research* **6**, pp. 773-782.

2.   Bosco C. and A. Carpinteri (1992): Fracture Mechanics Evaluation of Minimum Reinforcement in Concrete Structures. In: *Applications of fracture Mechanics to Reinforced Concrete*, ed. A. Carpinteri, Elsevier, London, pp. 347-377.

3.   Baluch M.H., A.K. Azad and W. Ashmawi (1992): Fracture Mechanics Application to Reinforced Concrete Members in Flexure. In: *Applications of Fracture Mechanics to Reinforced Concrete*, ed. A. Carpinteri, Elsevier, London, pp. 413-436.

4.   Hawkins N.M. K. and Hjorteset (1992): Minimum Reinforcement Requirements for Concrete Flexural members. In: *Applications of Fracture Mechanics to Reinforced Concrete*, ed. A. Carpinteri Elsevier London, pp. 379-412.

5.   Hededal O. and I. Kroon (1991): Lightly Reinforced High-Strength Concrete. M.Sc. Thesis in Civil Engineering, Department of Building Technology and Structural Engineering, Aalborg University, Denmark.

6.   Planas J., G. Ruiz and M. Elices (1995): Fracture of Lightly Reinforced Concrete Beams. In: *Theory and Experiments In Fracture Mechanics of Concrete Structures*, ed. F.H. Wittmann, Proc. of FRAMCOS-2 Conference, pp. 1179-1188.

7.   Petersson P.E. (1981): Crack Growth and Development of Fracture Zones in Concrete and Similar Materials. Ph.D. thesis, Lund Institute of Technology, Sweden.

8.   Brincker R. and H. Dahl (1989): Fictitious Crack Model of Concrete Fracture. *Magazine of Concrete Research* **41** ,No. 147, pp. 79-86.

APPENDIX C:

MULTIPLE CRACKING AND ROTATIONAL CAPACITY OF LIGHTLY REINFORCED BEAMS

Prepared by F.A. Christensen, M.S. Henriksen and R. Brincker

Abstract

In this appendix a model is formulated for the rotational capacity of reinforced concrete beams assuming rebar tension failure. The model is based on a classical approach and establishes the load-deflection curve of a reinforced concrete beam. The rotational capacity is then obtained as the area under the load-deflection curve divided by the yield moment of the beam. In calculating the load deflection curve, the cracking process of the concrete is ignored. By assuming that all cracks are fully opened, the energy dissipated during cracking of the concrete is taken into account by simply adding the total tensile fracture energy to the total plastic work obtained by the classical analysis.

Model Formulation

Before cracking of the concrete both the concrete and the reinforcement are assumed to behave elastically, and no slip is assumed between concrete and reinforcement. Assuming a linear variation of the normal beam strain over the cross-section, the stress distribution is obtained by classical beam theory.

When the tensile strength is reached at the tensile side of the beam, the concrete is assumed to crack. Further, cracks are assumed to be formed during constant bending moment (no decrease of the bending moment) and are allowed to extend until the level of the neutral axis. The tension force from the reinforcement is balanced by compression stresses in the concrete. The size of the compression zone is obtained by assuming a uniform distribution of the compression stresses and using an equilibrium equation. At the cracked section the tensile force in the reinforcement is transferred to the surrounding concrete by assuming a formation of two debonded zones around the crack with constant shear friction τ_f. In Figure C1 the stress distribution at a cracked section of the beam is shown immediately before and after formation of a crack.

Calculation Procedure

The crack development is initiated when the tensile strength is reached at the tensile side of the beam in the cross section with maximum bending moment. This corresponds to the situation shown in Figure C1. As the load increases, cracks might form in neighbour sections. If the bending moment is equal to the cracking moment at section II a new crack will be formed at section II. If the bending moment is less than the cracking moment, the load is increased causing the debonded zones to extend. By repeating this procedure cracks are formed one by one until tensile

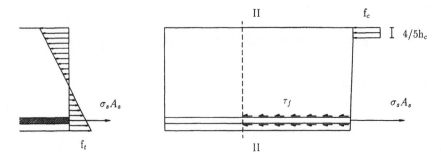

Figure C1. Stress distribution in cracked and uncracked cross-section. The debonded zone with constant shear friction stress ends at section II.

failure of the reinforcement bar. At the time a new crack is formed it is assumed that the strain at section II in Figure C1 is the same in the concrete and in the reinforcement and that the strain is equal to the tensile fracture strain of the concrete ($\epsilon = f_t/E_c$).

The load-deflection curve is found by integrating the curvature over the length of the beam. A typical load-deflection curve is shown in Figure C2. The curvature is determined as the ratio between the reinforcement strain and the distance from the reinforcement to the neutral axis at the cracked sections.

The rotational capacity is calculated by integrating the load deflection curve and adding the energy dissipated in the crack formation process estimated as nA_cG_F where n is the number of cracks, A_c is the cross sectional area of the beam and G_F is the fracture energy of the concrete.

Model Properties

Investigating the model properties, results have been derived using the following values for the material parameters:

$$
\begin{array}{rcl}
f_y & : & 500\ MPa \\
f_u & : & 575\ MPa \\
\tau_f & : & 5.0\ MPa \\
f_t\ (NSC) & : & 3.0\ MPa \\
f_t\ (HSC) & : & 5.0\ MPa \\
f_c\ (NSC) & : & 60.0\ MPa \\
f_c\ (HSC) & : & 90.0\ MPa \\
G_F & : & 0.120\ N/mm \\
\epsilon_y & : & 2.0\ \% \\
\epsilon_u & : & 15.0\ \% \\
E_s & : & 210,000\ MPa
\end{array}
$$

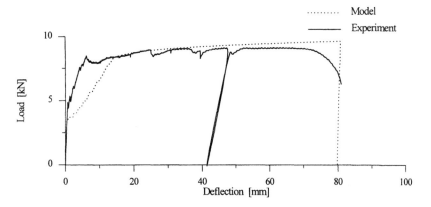

Figure C2. Failure response curve for $100 \times 200 \times 2400\ mm$ beam with 0.14 % reinforcement compared with typical experimental result.

representative for the two types of concrete used in the experimental investigation, see Appendix A.

For this kind of model, one of the most important parameters is the amount of reinforcement strain hardening described by the ratio f_u/f_y. If no strain hardening is present, i.e. if the ratio $f_u/f_y = 1$, at each crack, only one point (the point situated just between the concrete tensile failure crack faces) can be in the state of yielding. Thus, since the length of the zone over which yielding takes place tends to zero when the ratio tends to 1, the rotational capacity tends to zero. Furthermore, in this case only one crack will be formed reducing the possibilities of energy dissipation even further. The results clearly support these considerations. Figure C3 shows that the rotational capacity is highly dependent upon the ratio f_u/f_y.

One would expect the results of the model to be rather sensitive to the value of the shear friction stress τ_f. However, this is not the case, Figure C4. Generally it can be stated that the rotational capacity decreases with increasing friction stress, although the influence is small. The energy dissipation due to yielding and debonding decreases with increasing values of the shear friction stress, but at the same time the number of cracks increases, and thus, the contribution from dissipation of energy in the concrete tensile cracks increases.

Figure C5 shows the results for a normal strength concrete and a high strength concrete. As it appears from the figure, increasing the concrete strength decreases the rotational capacity for all values of the reinforcement ratio.

For the value of shear friction stress used here, $\tau_f = 5MPa$, the model only shows a small size effect. For larger values of τ_f, introducing a relatively larger contribution from concrete tensile fracture energy dissipation, somewhat larger size effects are observed, see the results in the main part of the chapter. However, the results in Figure C6 clearly show that the rotational capacity is non-sensitive to the size, but highly sensitive to the deformation capacity of the steel. For the low deformation capacity steel the value of the ultimate strain was reduced to $\epsilon_u = 2.0\%$, approximately equal to the failure strain of the cold deformed bars used in the experimental investigation, see Appendix A.

All Figures C3-C6 show how the rotational capacity is influenced by the reinforcement ratio. When reinforcement tensile failure controls the failure of the beam the rotational capacity is increasing with increasing reinforcement ratio.

Comparing with Experimental Results

Figure C7 and C8 show how the model compares with experimental results for the rotational capacity. In Figure C7 the case of normal strength concrete is shown.

In case of reinforcement ratios of 0.14% and 0.25% the model results fit the experiments well, whereas in some cases of 0.39% reinforcement ratio the experiments show that reinforcement tensile failure is no longer the dominating failure mode and therefore the model over-estimates the rotational capacity. The experimental results show, as the model, that the rotational capacity is higher in the case of normal strength concrete.

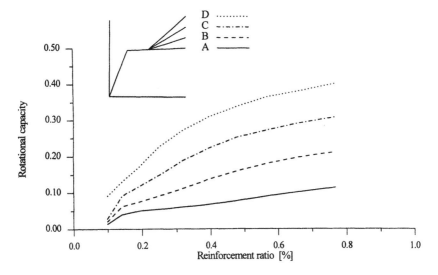

Figure C3. Rotational capacity as a function of the reinforcement ratio for different strain hardening properties A), B), C) and D) of the reinforcement.

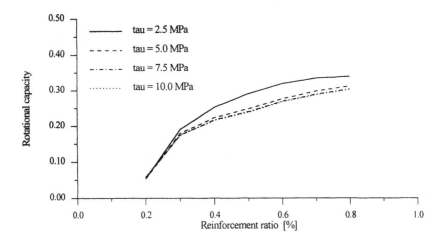

Figure C4. Rotational capacity as a function of the reinforcement ratio for different values of the debonding shear friction stress τ_f.

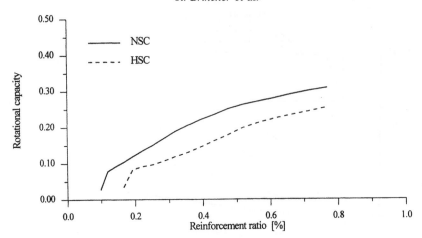

Figure C5. Rotational capacity as a function of the reinforcement ratio for normal strength concrete and high strength concrete.

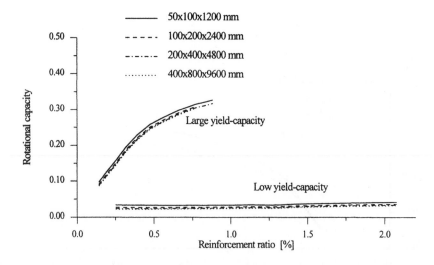

Figure C6. Size effect on the rotational capacity as a function of the reinforcement ratio for different reinforcement types: a low deformation capacity type, and a high deformation capacity type.

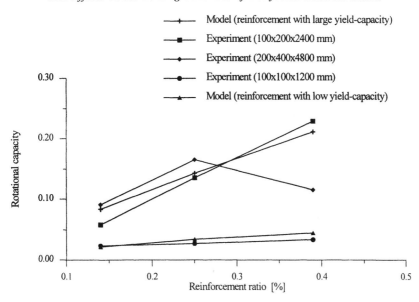

Figure C7. Comparison between rotational capacities obtained from model and from experiments (normal strength concrete).

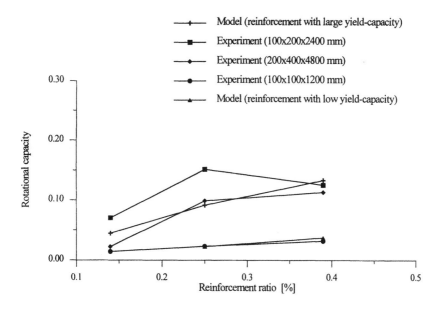

Figure C8. Comparison between rotational capacities obtained from model and from experiments (high strength concrete).

APPENDIX D:

FRACTURE MECHANICAL MODEL FOR ROTATIONAL CAPACITY OF HEAVILY REIN-
FORCED CONCRETE BEAMS

Prepared by M.S. Henriksen, R. Brincker and G. Heshe

Abstract

In this appendix the flexural behaviour of reinforced concrete beams is investigated by analytical
methods originally introduced by A. Hillerborg. A simple analytical model is presented which describes the bending moment-curvature relation for normal and over-reinforced beams taking into
account the strain localization within the compression zone of the concrete. The strain softening
part of the stress-strain curve for the concrete is described as a stress-deformation relation which
is dependent on the length over which the compression failure extends along the beam axis. On
the basis of the moment-curvature relation estimated by the model, the load-deflection curve is
calculated, and the rotational capacity is obtained as the total plastic work divided by the yield
moment of the beam. The results of the model are obtained assuming a linear compression softening curve with different values of the critical compression deformation and the fracture zone
length. The results are compared with experiments.

Introduction

In the past ten years a lot of research has been carried out in the field of compression failure
of reinforced concrete structures. After development of different crack models for the fracture
in tension like the Fictitious Crack Model, Hillerborg was one of the researchers that began to
investigate the softening behaviour of the fracture in compression [1],[2] and [3].

Hillerborg had already shown that the softening behaviour in tension was size-dependent, and in-
spired by the work of van Mier [4], Hillerborg got the idea of using the model for tension softening
on compression failure. This led to a simple model describing the uniaxial stress-strain relation
for concrete based on fracture mechanical concepts, Figure D1. The strain localization within
the compression zone was taken into account by defining a characteristic length dependent on the
depth of the compression zone.

Van Mier and Vonk at the Stevin Laboratory carried out different experimental studies on the full
range behaviour in compressive loading of different concrete cubes as well as performed micromechanical modelling of the compression softening [4],[5]. Recent studies of the behaviour in
compression of both normal strength concrete (45 MPa) and high strength concrete (90 MPa)
have been presented by Jansen and Shah [6], who performed experimental investigations on cylinders with constant diameter and different depths to examine the effect of specimen depth on
compressive strain softening of concrete, see Figure D2. Their results show that the post-peak

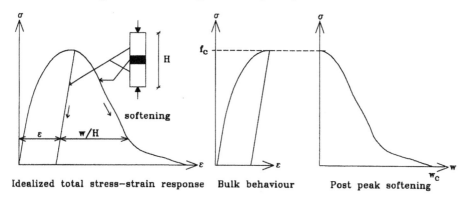

Idealized total stress–strain response   Bulk behaviour   Post peak softening

Figure D1. Basic idea of stress-strain relation in compression by Hillerborg [1].

behaviour including the post peak energy dissipation is relatively insensitive to the depth of the cylinder specimen.

Several researchers have examined the influence on compression softening behaviour when changing the depth of test cylinders and when using intermediate layers between the loading plate and the cylinder, but it seems to be a lack of investigations on the influence of changing the diameter. A so-called Compressive Damage Zone model has also been established by Markeset taking into account also localized shear deformation and deformation due to splitting cracks [7].

In this investigation, the length, over which the compression failure extends along the beam axis, is introduced as a characteristic length proportional to the depth of the compression zone $l_{ch} = \beta h_c$ and the softening is assumed to be linear. Thus, the model contains two parameters describing the basic fracture mechanical properties of the model: the characteristic length parameter β and a critical softening deformation w_c. In the following the influence of the parameters w_c and β on the full range behaviour of different model beams is analysed, and, based on the load-deflection curves, the rotational capacity is estimated as the total plastic work divided by the yield moment of the beam.

Basic Assumptions of the State of Compression Failure

When a reinforced concrete beam is loaded to ultimate compression failure and an unloading starts taking place, the critical cross-section is assumed to pass through three different states of failure, see Figure D3.

The continuum state for the critical cross-section describes an elastic state for the concrete where the concrete stresses $\sigma_c < f_c$ for all points in the cross-section. Varying the concrete strain ε_c from zero to the peak strain ε_0, the depth of the compression zone h_c will be constant. In this phase the reinforcement is assumed to be in an elastic state corresponding to $\sigma_s < f_y$.

The condition of fracture zone growth is satisfied when the concrete stress in the compressed edge

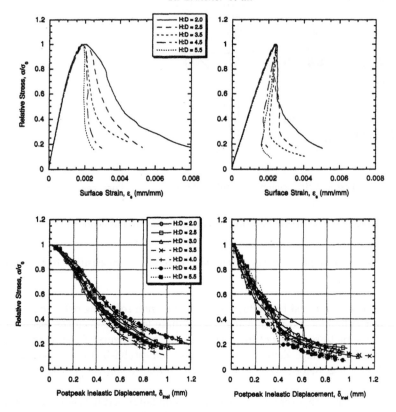

Figure D2. Influence of specimen length on concrete cylinder uniaxial compressive stress-strain curve, (diameter 100 mm). Left: Normal strength concrete (45 MPa). Right: High strength concrete (90 MPa). According to Jansen and Shah [6].

of the cross-section reaches the concrete compression strength. At this condition a fracture zone will start developing. When the fracture zone is fully developed, i.e. when the compression stress at the top of the beam has dropped to zero, the length of the fracture zone along the beam axis is assumed to be l_{ch}, where l_{ch} is defined as a characteristic length. The material within the fracture zone follows a softening branch and outside this zone an unloading takes place. The characteristic length could be assumed to be dependent on either the depth of the compression zone or on the width of the cross-section.

It is well known that the final compression failure of cylinders is often a so-called "cone-failure" where the concrete fails in a compression-shear mode with the development of slip-planes at an angle γ typically around $\gamma \cong 30°$. Thus, for a cylinder with radius r, the characteristic length might be defined as $l_{ch} \tan(\gamma) = 2r$, and then taking the approximation $\tan(\gamma) \cong 0.5$ we get $l_{ch} \cong 4r$, see Figure D4.

Following this idea, the failure mode of the compression zone of a beam is assumed to be a similar compression-shear mode with the development of slip-planes at a certain angle to horizontal. Now, assuming that the slip-planes will start at the point where the strain is zero, the characteristic length becomes proportional to the depth h_c of the compression zone, thus $l_{ch} = \beta h_c$. Assuming

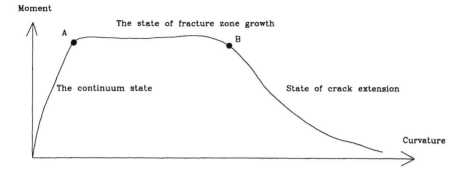

Figure D3. Full range behaviour for a reinforced concrete beam.

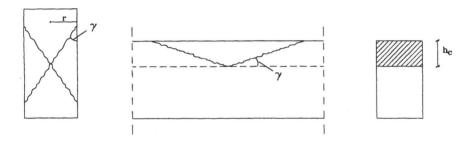

Figure D4. Definition of the parameter β describing the length of the failure zone. Left: Cone failure of a cylinder in compression. Right: Assumed failure mode in the compression zone of a beam in bending.

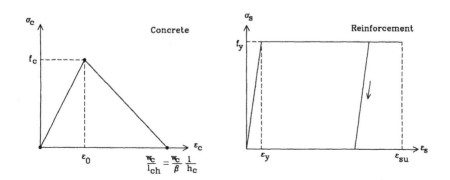

Figure D5. Simplified analytical stress-strain curves for the reinforcement and the concrete.

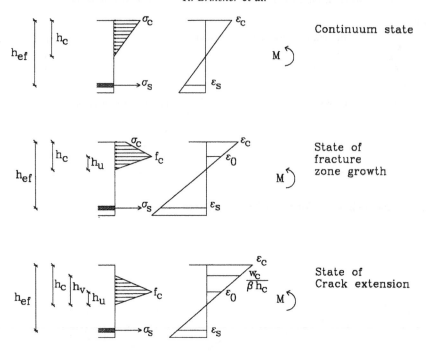

Figure D6. The stress and strain distribution for the critical cross-section.

that the slip-planes develop at an angle similar to the cylinder failure gives the estimate $\beta \cong 4$, Figure D4.

However, the relation $l_{ch} = \beta h_c$ is only expected to be valid on the assumption that the depth of the compression zone is small compared to the width b of the beam. If the beam width becomes substantially smaller than the depth of the compression zone, it is more reasonable to assume a failure mode where vertical slip-planes develop, and thus, for this case it should be assumed that the characteristic length is proportional to the width b of the beam. In the following however, the relation $l_{ch} = \beta h_c$ will be used.

Using the approach of a characteristic length l_{ch}, the softening deformation can be represented as a strain, and thus, the full range behaviour of both concrete and reinforcement can be represented by a stress-strain relation, Figure D5.

In the state of crack extension a part of the material in the compression zone has totally failed and a "real crack" is formed. The final failure develops as the crack extends downwards through the beam.

Modeling Flexural Behaviour

In the modeling it is assumed that the considered beams are subjected to three-point bending, and the critical cross-section is assumed to be reinforced only by main reinforcement. Thus, the influence of compressive reinforcement and stirrups is not taken into account. The bending tensile strength of the concrete is set equal to zero which means that the cross-section is assumed to be cracked from the start. Effects from other cracks along the beam axis as well as bond-slip effects between the concrete and the reinforcement are not considered. For calculation of the full range behaviour Bernuoilli's assumption (plane cross-sections remain plane) is applied.

The model describing the full range behaviour is based on simplified linear stress-strain curves for the reinforcement and the concrete, see Figure D5. Typical stress and strain distributions for the critical cross-section and each of the fracture states are shown in Figure D6.

The flexural behaviour is described for the three states as non-dimensional moment-curvature relations as follows:

The continuum state, $\epsilon_c \leq \epsilon_0$

$$\frac{6M}{bh_{ef}^2 f_c}(\xi, \epsilon_c) = \left(3\xi - \xi^2\right)\frac{\sigma_c}{f_c} \tag{1}$$

the state of fracture zone growth, $\epsilon_0 < \epsilon_c \leq w_c/\beta h_c$

$$\frac{6M}{bh_{ef}^2 f_c}(\xi, \epsilon_c) = 3\xi + \left(\frac{\epsilon_0}{\epsilon_c} - 2\right)\xi^2 + \left(\left(3 - 3\frac{\epsilon_0}{\epsilon_c}\right)\xi - \left(1 + \frac{\epsilon_0}{\epsilon_c}\left(\frac{\epsilon_0}{\epsilon_c} - 2\right)\right)\xi^2\right)\frac{\sigma_c}{f_c} \tag{2}$$

and the state of crack extension, $\epsilon_c > w_c/\beta h_c$

$$\frac{6M}{bh_{ef}^2 f_c}(\xi, \epsilon_c) = \left(3\frac{w_c}{\beta h_c \epsilon_c} + \frac{w_c^2}{(\beta h_c)^2 \epsilon_c^2}\right)\xi + \left(2\frac{w_c \epsilon_0}{\beta h_c \epsilon_c^2} - 3\frac{w_c}{\beta h_c \epsilon_c}\right)\xi^2 \tag{3}$$

where $\xi = h_c/h_{ef}$.

Bending moment-curvature curves and load-deflection curves are shown for different values of w_c and β in Figure D7. Here the curvature is calculated as the angle in the strain distribution ϵ_c/h_c. The results are shown for a reinforced normal strength concrete beam with reinforcement ratios (A_s/bh) 0.78 %, 1.57 %, 2.45 % and 4.02 %. Values for the beam dimensions are $b = 200$ mm, $h = 400$ mm, $l = 4800$ mm. The concrete parameters are $f_c = 60$ MPa, $\varepsilon_0 = 0.2$ % and the reinforcement parameters are $f_y = 600$ MPa, $E_s = 2.0 \times 10^5$ MPa, $\varepsilon_{su} = 10$ %. Note the large sensitivity of the modelled behaviour on the values of the two key parameters w_c and β.

Load-deflection curves are obtained by integrating the curvature distribution according to the principle of virtual work. When the curvature is integrated, it is kept constant over the characteristic length, see Figure D8 where the distribution of curvature along the beam axis is shown in the state of ultimate failure, here defined as the transition between the state of fracture zone growth and the state of crack extension.

The rotational capacity is estimated by integrating the load-deflection curves to obtain the total plastic work, and then dividing by the yield moment to obtain a non-dimensional parameter θ.

The parameter θ is a direct measure of the rotational capacity of the beam.

Results for the rotational capacity are shown in Figures D9-D12 for different values of w_c and β. As expected, the rotational capacity is strongly sensitive to both parameters. Note that since the ultimate strain of the reinforcement is incorporated in the analysis, some of the test results correspond to reinforcement tensile failure. The reason for incorporating this rather simple tension failure in this model (no bond-slip is modeled) is to have a rough check on the capability of the model to show a reasonable switch between the two modes of failure. The part of the rotational capacity curves in Figures D9-D12, where the rotational capacity is increasing with the reinforcement ratio, corresponds to reinforcement tension failure. This part of the results should be acknowledged as less interesting (modelling for this part is given in Appendix C) than the rest of the curve corresponding to concrete compression failure.

More details of the modelling are given in Henriksen and Brincker [8].

Comparison with Experimental Results

The values of the rotational capacity estimated by the model are shown in Figures D13-D16 as a function of the reinforcement ratio and compared with experimental results from Appendix A for normal strength concrete beams with a cross-section of $100 \times 100\ mm$, $100 \times 200\ mm$ and $200 \times 400\ mm$ and slenderness number 12. As it appears, the best agreement is obtained for the values $w_c = 4\ mm$ and $\beta = 8$.

Figure D16 shows that using the values $w_c = 4\ mm$ and $\beta = 8$, the rotational capacity estimated by the model compares reasonably well with experimental results for all the three beam sizes.

References

1.   Hillerborg, A. (1990): Fracture Mechanics Concepts Applied to Moment Capacity and Rotational Capacity of Reinforced Concrete Beams. *Engineering Fracture Mechanics* **35**, No. 1/2/3, pp. 233-240.

2.   Hillerborg, A. (1988): Rotational Capacity of Reinforced Concrete Beams. *Nordic Concrete Research* **7**, pp. 121-134.

3.   Hillerborg, A. (1991): Size Dependency of the Stress-Strain Curve in Compression. In: *Analysis of Concrete Structures by Fracture Mechanics*, Rilem Report 6, eds. L. Elfgren & S. P. Shah, Chapman & Hall, London, 171-178.

4.   van Mier, J.G.M. (1986): Multi-axial Strain-Softening of Concrete. *Materials and Structures* **19**, No. 111, Rilem, pp. 179-200.

5.   Vonk, R. A. (1993). A Micromechanical Investigation of Softening of Concrete loaded in Compression. *Heron* **38**, No. 3.

6.   Jansen, D.C. and S.F. Shah (1997): Effect of Length on Compressive Strain Softening of Concrete. *Journal of Engineering Mechanics* **123**, No. 1, pp. 25-35.

7.   Markeset, G. (1995): A Compressive Softening Model for Concrete. In: *Fracture Mechanics of Concrete Structures*, Proceedings FRAMCOS-2, edited by Folker H. Wittmann.

8.   Henriksen, M.S. and R. Brincker (1998): Modelling the Rotational Capacity of Heavily Reinforced Concrete Beams. To be published in Nordic Concrete Research, Publication No. 21.

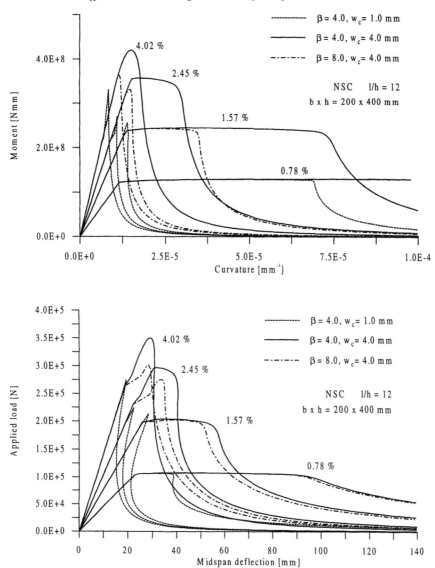

Figure D7. Model results for moment-curvature (top) and load-deflection (bottom) relation for a normal strength concrete beam ($200 \times 400 \times 4800 \ mm$) and different values of w_c and β.

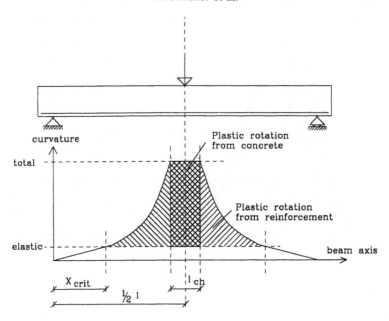

Figure D8. Typical curvature distribution along the beam axis for a beam at ultimate load.

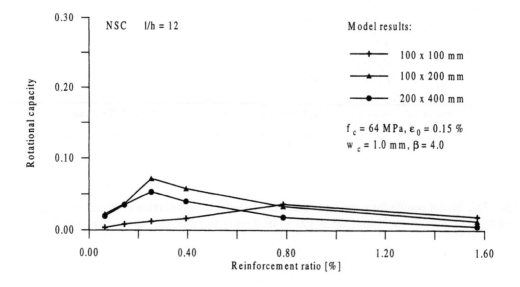

Figure D9. Model results for normal strength concrete beams with $w_c = 1.0$ *mm* and $\beta = 4.0$.

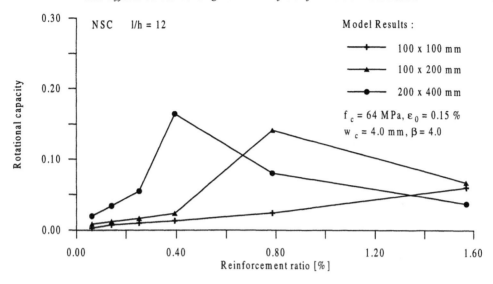

Figure D10. Model results for normal strength concrete beams with $w_c = 4.0\ mm$ and $\beta = 4.0$.

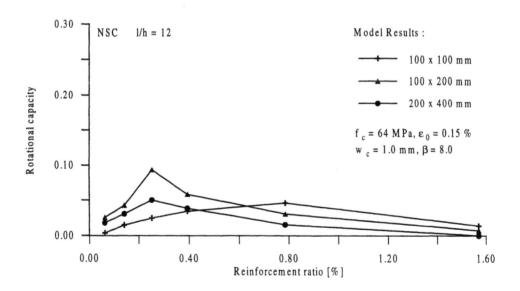

Figure D11. Model results for normal strength concrete beams with $w_c = 1.0\ mm$ and $\beta = 8.0$.

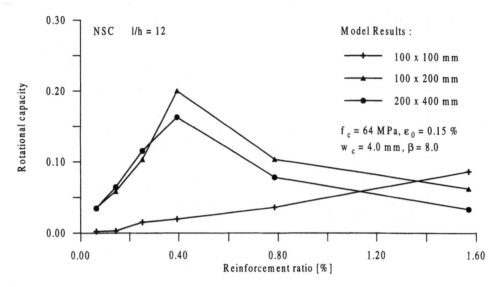

Figure D12. Model results for normal strength concrete beams with $w_c = 4.0\ mm$ and $\beta = 8.0$.

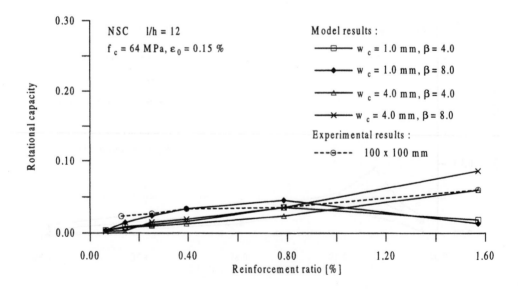

Figure D13. Results for $100 \times 100 \times 1200\ mm$ normal strength concrete beams.

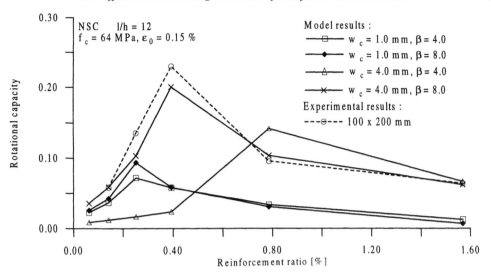

Figure D14. Results for $100 \times 200 \times 2400$ mm normal strength concrete beams.

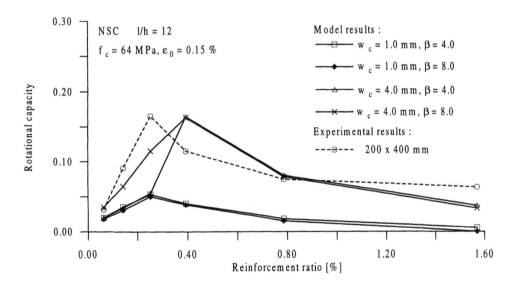

Figure D15. Results for $200 \times 400 \times 4800$ mm normal strength concrete beams.

R. Brincker et al.

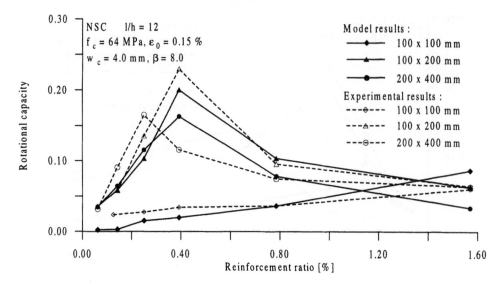

Figure D16. Results for normal strength concrete beams with $w_c = 4.0$ *mm* and $\beta = 8.0$.

MINIMUM REINFORCEMENT REQUIREMENT FOR RC BEAMS

J. OZBOLT and M. BRUCKNER
*Institut für Werkstoffe im Bauwesen, Universität Stuttgart,
Pfaffenwaldring 4, D-70550 Stuttgart, Germany*

ABSTRACT

In the present paper different aspects of the requirement for the minimum reinforcement ratio are studied and discussed. The influence of the beam depth is investigated in more detail. Numerical analysis for reinforced concrete beams of different sizes is carried out using plane finite element code *MASA2* which is based on the nonlocal mixed constrained microplane model. Presently, an extensive test project for reinforced concrete beams in which the material and geometrical properties are varied is in progress. Currently available test results are compared with the numerical results. It is concluded that the requirement on the minimum reinforcement depends on the beam size but also on the material properties as well as on the amount and type of the distributed reinforcement. To define the dependency between the minimum reinforcement and geometrical as well as material parameters in more detail, further theoretical and experimental studies are needed.

KEYWORDS

Minimum reinforcement, RC beams, nonlocal microplane model, energy criteria.

INTRODUCTION

In engineering practice RC beams of different sizes and with different reinforcement ratios are often used. They are normally designed such that the internal forces as well as their distribution over the cross section are calculated according to the elastic beam theory. On the contrary, the dimensioning is performed using a limit state procedure. Obviously, this is in contradiction. Therefore, significant efforts have recently been made in order to develop consistent tools and recommendations for the nonlinear structural analysis and dimensioning according to the limit state procedure. In order to provide enough structural safety and to make a redistribution of internal forces possible, RC beams must be designed such that they fail in a ductile manner. Some recent fracture mechanics studies [1-5] indicated that larger beams are more brittle than smaller. Consequently, the question is whether the same material laws and design rules may be used for RC beams of different sizes.

An important condition for ductile failure of RC beams is the minimum reinforcement requirement. The minimum reinforcement must assure a stable and ductile beam response after the concrete tensile strength at the beam tensile zone is reached. Presently, in almost all design codes the minimum reinforcement requirement is independent of the beam depth. For example, according to [6] the typical value .for the minimum reinforcement ratio is $\rho_{min} = 0.14\%$ ($f_c = 32$ MPa). Recent theoretical and experimental investigations [1,3,7], which introduced fracture mechanics into the consideration, indicated that the minimum reinforcement area must depend on the beam size.

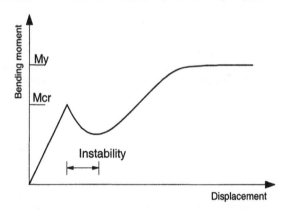

Fig. 1. Unstable load-displacement response after reaching M_{cr}.

Following the equilibrium condition criteria (stress stability condition), brittle failure occurs then when $M_{cr} \geq M_y$ with M_{cr} = bending moment at opening of the first bending crack and M_y = bending moment due to the yielding of reinforcement (see Fig. 1). According to the aforementioned work, to assure a stable beam response the minimum reinforcement requirement should be calculated employing the stress stability criteria [7]:

$$M_{cr} = M_y \tag{1}$$

For proportionally scaled beams Eq.(1) leads to smaller minimum reinforcement ratio for larger beams. This is due to the fact that with increase of the beam size, M_{cr} decreases as a consequence of the size effect on the nominal bending strength. Therefore, assuming no size effect on M_y it follows that the minimum reinforcement area decreases when the beam size increases i.e. according to [7]:

$$\rho_{min} = N_c / (f_y \sqrt{h}) \tag{2}$$

with N_c = constant depending on the concrete properties, f_y = reinforcement yield stress and h = beam depth. Obviously, according to Eq. (2) for very large beams the minimum reinforcement ratio yields to zero.

In contrast to theoretical studies, the experimental work in [8] indicated that the minimum reinforcement ratio is related to the type of reinforcement rather than to the size of the beam. Anyhow, in spite of intensive research in this field the minimum reinforcement requirement is still an open question and, therefore, a subject of intensive studies.

To better understand the above mentioned problems a numerical fracture finite element analysis for geometrically similar RC beams of different sizes and with different reinforcement ratios is carried out. The analysis is performed by the use of the nonlocal microplane model [9,10] using the plane finite element code *MASA2*. Furthermore, the numerical results are compared with the available test results which are a part of an extensive test program that is still in progress.

MECHANICAL CONSIDERATIONS RELATED TO THE MINIMUM REINFORCEMENT REQUIREMENT

Let us first define the criteria for the minimum reinforcement. The minimum reinforcement is defined as a bending reinforcement concentrated at the bottom of the beam which must assure: **1.** Stable beam response after M_{ft} is reached (M_{ft} corresponds to the bending moment at which the tensile strength of concrete at the beam bottom is reached) and **2.** the ultimate bending moment M_u must be approximately equal or larger than M_{ft}. In addition to the above criteria, the minimum reinforcement ratio should not be to high since otherwise the beam may fail in diagonal shear or concrete compression rather than in bending.

Let us first consider the minimum reinforcement ratio for RC beams without stirrups and with any distributed reinforcement, which is the most extreme case occurring in practice. As aforementioned, the minimum reinforcement requirement has recently been extensively investigated by Carpinteri and co-workers. According to their work the minimum reinforcement ratio should decrease when the beam depth increases. However, from the theory [11,12] and experiments [2] we know that large plain concrete beams generally exhibit more brittle (explosive) behavior after cracks in concrete occur than the small beams. Comparing the response of a small plain concrete beam with an extremely large one, it can be observed that the small beam exhibits a relatively ductile response, i.e. even without any reinforcement the beam response is relatively stable. Consequently, for small beams the energy equilibrium criterion is not relevant and therefore for the minimum reinforcement requirement we propose the stress equilibrium criterion in the form:

$$M_u = M_y + M_{con} \geq M_{f_t} \tag{3}$$

were M_{con} stays for the contribution of concrete to the ultimate resistance. Note that there is an important difference between Eq. (1) and Eq. (2) proposed by Carpinteri and Eq. (3). According to (1) $M_y \geq M_{cr}$, which leads to the increase of the minimum reinforcement when the beam size increases. Since the minimum reinforcement should prevent brittle failure, there is no much sense to increase it if the structural response anyhow tends to be more ductile by decreasing the beam size. This is the main reason why we propose a stress equilibrium criterion different than that proposed by Eq. (1).

From the stress equilibrium point of view, Eq. (3), the most critical case occurs when the contribution of concrete to the ultimate resistance is small, i.e. $M_{con} = 0$ (for instance the concrete fracture energy $G_F \rightarrow 0$, as for example in pre-cracked concrete or in very large beams). Therefore, from Eq. (3) it follows:

$$M_y \geq M_{f_t} \quad \Rightarrow \quad \rho_{\min} \geq \frac{f_t h}{6 f_y z} \tag{4}$$

were f_y = yield limit of the reinforcement, z = lever arm between tensile and compressive forces of the cross-section. The above criterion is independent of the beam size. It is the lowest practical minimum reinforcement ratio because the energy criterion becomes relevant when a brittle cracking occurs, i.e. for larger beams. Theoretically, for $h => 0$ one needs no minimum reinforcement since the concrete is for such a case perfectly elasto-plastic already without any reinforcement (for $h \rightarrow 0$, $(\partial U / \partial a) / G_F \rightarrow 0$; with U = total accumulated energy, a = crack length).

In contrast to small beams, a large plain concrete beam exhibits an elasto-brittle response and fails immediately after the peak load (tensile strength of concrete) is reached. Therefore, intuitively, one would expect that large beams need relatively more bending reinforcement in order to prevent such an explosive response. The reason for the explosive behavior of large plain concrete beams after the first bending crack initiates is caused by relatively large energy release rate compared to the concrete fracture energy. To assure a stable structural response one needs reinforcement which has to take up the forces and to consume the energy released as a consequence of cracking. Assuming geometrical similarity, the structural energy release increases approximately proportionally with the square root function of the beam size. In order to fulfill the energy equilibrium after the crack starts to grow, for a constant reinforcement ratio the reinforcement strains must also increase with the same proportion. This is possible only when the cross-section after crack initiation remains approximately plane (bending theory applies), i.e. when the beam cross-section rotates proportionally with the crack growth. As will be shown later, for small reinforcement ratios this is valid only for relatively small beams. For larger beams, however, the cross-section generally does not remain plane i.e. due to the large energy release rate in a large beam with a small reinforcement ratio, the bending crack can propagate almost without any rotation of the beam cross-section.

Another important phenomenon in large beams is the anchorage of bending reinforcement. The reinforcement can be effectively utilized if an effective anchorage is possible. With increase in the beam size the tensile force in the reinforcement increases as well. However, due to the higher reinforcement density in the bottom beam zone (2D similarity), the anchorage resistance area (volume) increases less than proportionally. Consequently, the anchorage capacity relatively decreases when the beam size increases. This means that in large beams the critical cross-section must rotate more than proportional in order to activate the steel bars with the same efficiency. Obviously, the minimum reinforcement ratio should also be a function of the reinforcement type, the amount of the distributed reinforcement and the stirrups which strongly influence the anchorage capacity as well as the propagation of the bending cracks into the beam compressive zone.

Regarding the discussion above, Eq. (3) is only one of two criteria. To assure a stable structural response for large beams after the concrete tensile strength at the beam bottom is reached, the energy equilibrium should also be fulfilled. This criterion may be qualitatively formulated as:

$$(\Delta U / \Delta a)_{f_t} < G_R \qquad\qquad (5)$$

where $(\Delta U / \Delta a)_{ft}$ is the structural energy release rate when M_{ft} is reached and G_R = energy consuming rate by reinforcement and concrete in the vicinity of the reinforcing bars (bond-slip). If $G_R < (\Delta U / \Delta a)_{ft}$ unstable crack propagation occurs even if $M_y > M_{ft}$. Therefore, in contrast to small beams, for large beams Eq.(5) governs the minimum reinforcement area since the energy accumulated in the beam is large. Consequently, the bending reinforcement in large beams is less effective and one needs relatively more of it to assure a stable beam response.

NUMERICAL ANALYSIS

To investigate the influence of the reinforcement on the response of RC beams of different sizes and with different amount of reinforcement, numerical analysis was carried out [13]. In the analysis the finite element code based on the nonlocal microplane model for concrete and two-dimensional finite elements were used. First, the influence of concrete on the ultimate resistance and load-displacement response for beams of different sizes which fail by yielding of reinforcement were investigated. Subsequently, the minimum reinforcement was studied in more detail.

Geometry and Material Properties

The numerical study was carried out for reinforced concrete beams loaded in three-point bending. Geometrically similar beams of five different sizes (h = 100, 200, 400, 800 and 1600 mm; see Fig. 2) were analyzed. The beam width (b = 100 mm) as well as the span-beam depth ratio (l/h = 6) were kept constant. In the study the reinforcement ratio (ρ= $100 \cdot A_s / (b \cdot h)$; A_s = reinforcement area) was varied from 0 to 2% for each beam size. The bending reinforcement is assumed to be placed at the beam bottom with $d = 0.9h$ (see Fig. 2).

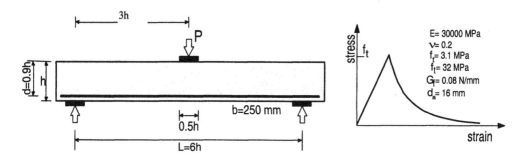

Fig. 2. Geometry of the beam and local tensile stress-strain curve.

The concrete properties employed in the analysis were constant for all beam sizes and taken as follows: uniaxial tensile strength f_t = 3.1 MPa, uniaxial compressive strength f_c = 32 MPa, fracture energy G_F = 0.08 N/mm and maximum aggregate size d_a = 16 mm. The shape of the adopted local tensile stress-strain softening curve is shown in Fig. 2. An ideally elasto-plastic stress-strain relationship for steel is adopted with Young's modulus E_S = 210000 MPa and yield limit f_y = 420 MPa. The reinforcement is introduced in a smeared sense, i.e. as a layer(s) of elements in the bottom zone of the beam. Although a bond-slip relationship between steel and concrete is not explicitly specified, it is taken into account in an integral form through the rows of finite elements below and above the reinforcement. These rows of elements can be interpreted as bond-slip interface elements. The properties of the material model (microplane material model) are calibrated such that the objectivity of the concrete shear resistance close to the reinforcement is assured. Since the analysis is nonlocal the size effect on the bond-slip resistance is automatically taken into account. The load was applied by controlling the vertical displacement under the loading plate. To take into account the confinement effect of the loading plate, the displacements under the plate in horizontal (beam span) direction were kept fixed and equal to zero. The dead load of the beam is not taken into account.

Contribution of Concrete to the Ultimate Bending Moment

In Fig. 3 the typical load-displacement curves (L-D) for small and large beams with low reinforcement ratio are shown. It can be seen that the small beam exhibits a rather ductile response. On the contrary large beam fails after crack initiation. To demonstrate the contribution of concrete to the peak resistance of beams with relatively small reinforcement ratio (ρ = 0.25 - 0.375 %) which fail due to the yielding of reinforcement, the nominal bending moment that corresponds to the peak load (plotted as a function of the beam depth) is shown in Fig. 4. The bending moment is normalized to M_y i.e. the lowest possible bending (yielding) moment is calculated according to:

$$M_y = \sigma_y A_s (0.9h)$$

(6)

with $d = 0.9h$ = effective beam depth. Eq. (6) is used in design practice and it neglects the contribution of concrete to the peak load. Fig. 4a shows that small RC beams with low reinforcement ratio have much higher peak resistance than the ultimate resistance predicted by Eq. (6), i.e. for these beams the contribution of concrete to the peak resistance is approximately the same as the contribution of reinforcement. This contribution, compared with the contribution of reinforcement (small reinforcement ratio), is relatively large and it is a consequence of the stable crack growth in small beams.

When the beam size increases the contribution of concrete decreases (see Fig. 4) and for very large beams the maximal bending moment is controlled only by the reinforcement. Note that the numerical results probably show slightly too high values for the contribution of concrete to the peak resistance. The reason is due to the relatively high assumed tensile strength of concrete and relatively low yield stress of reinforcement (420 MPa). Furthermore, in the analysis the hardening of the reinforcement was neglected. Beams with a low reinforcement ratio fail by rapture of steel. Therefore, depending on the steel hardening ratio the failure moment will increase and the relative contribution of concrete at peak load should decrease. However, the present numerical results, which have been obtained before the experimental program was carried out, qualitatively clearly show that in small beams with low reinforcement ratio the contribution of the concrete is significant.

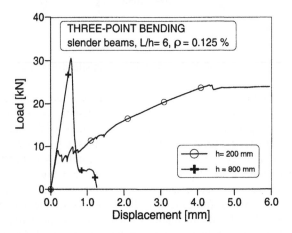

Fig. 3. Typical load-displacement curves for beams of different sizes with low reinforcement ratio.

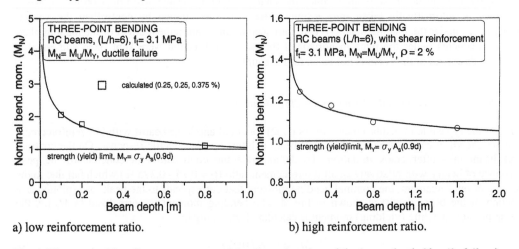

a) low reinforcement ratio.   b) high reinforcement ratio.

Fig. 4. The nominal bending moment at peak load as a function of the beam depth (ductile failure).

Ductile failure of RC beams with higher reinforcement ratio (RC beams with $\rho = 2\%$) was investigated for the size range $h = 100$ to 1600 mm. To prevent diagonal shear failure the beams were reinforced by stirrups ($\phi10/100$ mm). The beams were optionally provided by distributed reinforcement as shown in Fig. 5. The material behavior of this reinforcement is assumed to be ideally elasto-plastic with the same yield limit as for the bending reinforcement (420 MPa).

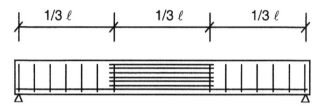

Fig. 5. Stirrups and distributed reinforcement.

In Fig. 4b the nominal bending moment at peak load is plotted as a function of the beam size. As can be seen, in contrast to the RC beams with relatively small reinforcement ratio, the size effect on the nominal bending moment is not significant. Namely, by doubling the beam depth (from 400 to 800 mm), the nominal strength decreases only by about 5%. The reason for relatively small size effect in beams with higher reinforcement ratio is due to small contribution of the concrete tensile resistance.

Minimum Reinforcement

The numerical results show that for relatively small beams ($h = 100$ and 200 mm) the sufficient minimum reinforcement ratio is approximately 0.125 %. When in these beams the bending moment reaches M_u, the L-D curve exhibits an horizontal plateau (see Fig. 6), the reinforcement yields and the beam shows a stable and ductile response. The energy stability criterion is fulfilled ($(\Delta U / \Delta a)_{fl} < G_R$) since for small beams the structural energy release rate is relatively small.

Obviously, for small beams the energy criterion is not relevant since for $h \to 0$, the beam response tends to be ideally elasto-plastic, i.e. ductile. Furthermore, the numerical results show that in small beams with low reinforcement ratio the contribution of the concrete to the ultimate resistance is high and the response is ductile. The analysis confirmed that for these beams the stress stability criterion is relevant. According to Eq. (4) the required minimum reinforcement ratio for the given geometry and material properties leads to 0.15%, which is in good agreement with the numerical results.

When the beam size increases up to $h = 800$ mm and the percentage of reinforcement is kept constant ($\rho = 0.125\%$) a clear tendency for an unstable beam response after M_{fl} is reached can be observed. In order to show this more clearly the normalized load-displacement curves plotted for beam depths varied from $h = 100$ to 800 mm with a low and constant percentage of reinforcement ($\rho = 0.125\%$) are shown in Fig. 6. It can be seen that the beam sizes up to approximately $h = 200$ mm, after a critical unstable point (the relative load = 1), exhibit a stable response. The energy stability condition, Eq. (5), is satisfied and further stable response is assured. On the contrary, larger beams ($h = 800$ mm, $\rho = 0.125\%$) fail at crack initiation in a brittle manner. Namely, after reaching M_{fl} the energy stability condition $(\Delta U / \Delta a)_{fl} < G_R$ is not reached before fulfilling the stress stability condition $M_u = M_y + M_{con} \geq M_{fl}$. The beam fails before the reinforcement resistance capacity can be activated. The first bending crack runs into the compression zone with almost no rotation of the critical cross-section.

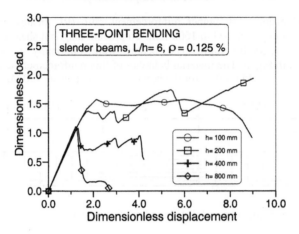

Fig. 6. Calculated load-displacement curves for different beam sizes and constant (small) reinforcement ratio (ρ = 0.125%). The reference values correspond to displacements and bending moments calculated according to the elastic bending theory for $\sigma_{max} = f_t$.

Due to energy reasons and to the fact that the bending reinforcement is localized at the beam bottom, the relatively low reinforcement ratio (ρ = 0.125%) can not prevent unstable crack growth in a large beam. Consequently, deformed cross-section of the beam does not remain plane. To confirm this, in Fig. 7 the distribution of the strains over the beam mid-span cross-section are plotted for a small beam (h = 100 mm, ρ = 0.125) and for a large beam (h = 800 mm, ρ = 0.125), close after the first bending crack initiates. As can be seen, for the large beam the position of the neutral axis is much closer to the upper compression beam edge than in the small beam. Actually, in the large beam almost the whole part of the cross-section is loaded in tension and the beam fails in shear close to the edge of the loading plate. The strain distribution over the mid-span cross-section (see Fig. 7) shows a strong nonlinearity, i.e. the hypotheses of the plane cross-section obviously does not hold. Consequently, the reinforcement area in large beam is effectively used only locally (at the beam bottom), i.e. it can not prevent no-rotational crack growth due to the relatively high energy release rate. In contrast to a large beam, the strain distribution over a cross section for a small beam is close to the bending theory (see Fig.7), i.e. the deformed cross-section remains approximately plane. The steel resistance capacity can be effectively used since the beam critical cross-section rotates approximately proportionally with the crack growth.

Unfortunately, presently no systematic tests for very large RC beams with low percentage of reinforcement exist. Bosco and Carpinteri [2] performed tests on beams of the same geometry as used in the present study. The dimensionless load-rotation curves measured in experiments, for beams with h = 100, 200 and 400 mm and with ρ_{min} obtained from Eq. (2), are plotted in Fig. 9. The figure demonstrates that when the beam depth increases from h = 100 mm (Fig. 8, case A with ρ_{min} = 0.256 %) to h = 400 mm (case C with ρ_{min} = 0.128 %) the stability of the response decreases, i.e. already for a relative small beam depth (h = 400 mm) the load-rotation curve indicates a decrease of the load after M_u is reached by approximately 20 %, before a stable response and M_y can be activated. Transferring these results gained from a displacement controlled test to the loading conditions of real structures, the drop of the load after reaching the ultimate moment may cause a brittle failure. To prove this in more detail further experiments on large beams are needed.

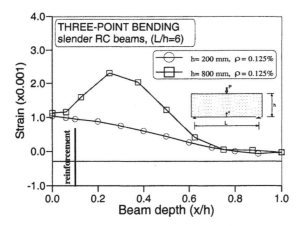

Fig. 7. Distribution of axial strains in the mid-span beam cross-section close after reaching M_u for a small beam (h = 100 mm) and for a large beam (h = 800 mm), ρ = const = 0.125 %.

As pointed out at the beginning of this section, the above discussed criteria for the minimum bending reinforcement ratio are very much on the safe side since in the practice bending reinforcement comes always together with stirrups and sometimes with horizontal distributed reinforcement. To demonstrate the importance and efficiency of the distributed reinforcement on the required minimum amount of bending reinforcement in larger beams, a beam with h = 1600 mm and ρ = 0.14 % (which is the minimum reinforcement ratio according to CEB-FIP [6] recommendations) is analyzed with and without distributed reinforcement. The amount of distributed reinforcement is assumed to be 0.1% of the total cross-section area distributed over the bottom half of the beam depth. It is modeled in a smeared sense, in horizontal and vertical direction (stirrups), assuming the same yield limit as for the bending reinforcement (f_y = 420 MPa).

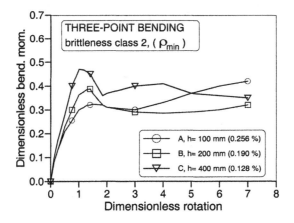

Fig. 8. Dimensionless load-rotation response for beam depths h = 100, 200 and 400 mm obtained in experiments [2] using ρ_{min} calculated according to Eq. (1).

In Fig. 9 the calculated L-D curves for both cases are plotted. The figure shows that 0.14% of the minimum bending reinforcement together with the distributed reinforcement of 0.1% assures a stable response after M_u is reached. On the contrary, as already shown before, the reinforcement ratio of

0.14% with no structural reinforcement is lower than the required minimum (approximately 0.5%). The total amount of minimum reinforcement, $\rho_{min} = \rho_{bend} + \rho_{dis} = 0.14 + 0.10 = 0.24$ %, is about a half of what we would need if there was no distributed reinforcement.

Obviously, in large beams the distributed reinforcement effectively contributes to the stability of the response. The reason is due to the fact that the damage caused by concrete cracking is distributed over a larger concrete volume which makes the consumption of the structural energy release possible. The minimum amount of the distributed reinforcement should be a function of the beam size. For the present example ($h = 1600$ mm) the assumed minimum reinforcement ratio of 0.1% seems to be sufficient. In contrast to large beams, in small beams one needs practically no distributed reinforcement.

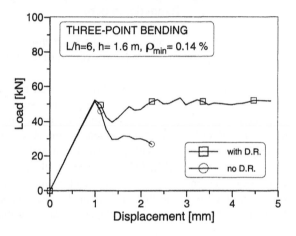

Fig. 9. Calculated load-displacement curves for large RC beam ($h = 1600$ mm) with $\rho_{min} = 0.14\%$ of bending reinforcement, with and without distributed reinforcement.

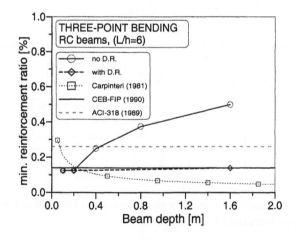

Fig. 10. Calculated minimum reinforcement area as a function of the beam depth obtained in the numerical study compared with [6] ($\rho_{min} = 0.14\%$), [15] ($\rho_{min} = 0.26\%$) and [7].

In Fig. 11 the minimum reinforcement ratio for RC beams without and with structural reinforcement is plotted as a function of the beam depth. In RC beams without structural reinforcement that are smaller than $h = 200$ mm the minimum reinforcement ratio obtained in the numerical study may be taken as a constant ($\rho_{min} \approx 0.14\%$) and it goes along with most of the current recommendations. However, for beams larger than $h = 200$ mm the minimum reinforcement ratio increases from 0.14% to approximately 0.5% for a beam depth of $h = 1600$ mm and is approximately a square root function of the beam depth. Therefore, when doubling the beam depth the minimum reinforcement ratio increases by about 40%. This result may be qualitatively confirmed by the single crack Jenq-Shah model [14] which is based on the *LEFM*. According to the model, when the bending crack initiates the contribution of the reinforcement to the resistance is a square root function of the ratio between the reinforcement area and the beam depth. This means that the reinforcement area must increase proportionally with the square root of the beam depth in order to contribute proportionally to the ultimate load and to prevent failure at crack initiation.

In contrast to large RC beams without distributed reinforcement, Fig. 10 indicates that in beams with a certain 'minimum' amount of distributed reinforcement (assumed 0.1%) the minimum bending reinforcement practically does not increase when the beam depth increases.

In Fig. 11 the numerical results for the minimum reinforcement ratio are also compared with the equation proposed in [7], which is based on the stress stability criterion, Eq. (1), that includes the size effect on the concrete bending strength. For comparison, the provisions by [15] and CEB-FIP design codes [6] are plotted for the same concrete properties as assumed in the present numerical study. For the beam without distributed reinforcement the amount of the minimum bending reinforcement ratio obtained from the analysis indicates disagreement with the code provisions. On the contrary, for the beams with a certain minimum amount of distributed reinforcement, which should be related to the reinforcement type and beam size, the results of the numerical study indicate good agreement with the design code formulas for the whole size range. The results of the analysis, however, show disagreement with the theoretical predictions based on the stress stability criterion (see Eq. (1)) proposed in [3].

EXPERIMENTAL RESULTS

Introduction

In order to get information about the structural behaviour of low reinforced concrete beams of different sizes, test were carried out at the Institut für Werkstoffe im Bauwesen, Univerität Stuttgart. The current experimental results are briefly presented here and compared with the usual analysis according to [16]. The reinforcement was designed as close as possible to fulfil the field-situation with a minimum reinforcement ratio of 0.15% according to [16]. To be very close to the practical conditions, stirrups and reinforcement in the compression zone of the concrete were used.

Geometry

Three different sizes of beams were cast. Their geometries are shown in Fig. 11.

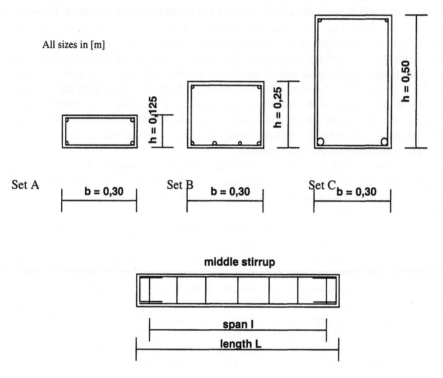

Fig 11. Geometry and reinforcement of beams.

The reinforcement ratio was chosen as low as possible according to [16]:

$$A_s / A_c = r_l = 0.15\%$$ (7)

The relation between the span and the beam depth was chosen $l/h = 6$, the length to the depth was chosen $L/h = 7$. An overview of the tested geometries is given in Table 1.

Table 1. Beam geometries

Set	h [m]	b [m]	l [m]	L [m]	ρ_l [%]	reinforcement	stirrups	spacing [cm]
A	0.125	0.30	0.75	0.875	0.15	2 ϕ 6	ϕ 6	s = 12.5
B	0.250	0.30	1.50	1.750	0.15	4 ϕ 6	ϕ 6	s = 15.0
C	0.500	0.30	3.00	3.500	0.15	2 ϕ 12	ϕ 6	s = 20.0

In order to initiate the major crack in the middle of the beam one stirrup was arranged exactly in the vertical symmetry axis of the beam. The spacing of the stirrups was taken according to the minimum reinforcement requirements [16] and computed to avoid shear failure of the beams. The concrete cover was not scaled and it was fixed to $c = 30$ mm. To guarantee the concrete cover thickness plastic bar spacers were used.

Material Properties

Concrete. The normal strength concrete C25/30 was used. The maximum aggregate size was 16 mm. The specimens were cast in a company for precasting concrete members. They were kept in the formwork for 20 days and watered every day in order to avoid cracks due to shrinkage of concrete. Table 2 shows the compressive cube strength, splitting tensile strength and bending tensile strength of the specimens A, B and C. The splitting and the bending tensile strength were calculated according to [6]. The bending tests were carried out on 4- point-bending test specimens.

Table 2. Concrete strength

Set	Compressive cube strength [N/mm²]	Splitting tensile strength [N/mm²]	Bending tensile strength [N/mm²]	f_{ctm} [N/mm²]
A	31.80	3.03	3.07	2.73
B	31.80	3.03	3.07	2.73
C	33.60	3.13	3.18	2.73

To obtain a realistic value of the fracture energy that can be used in the Finite Element analysis, tests according to [17] were carried out. The results are given in Table 3.

Table 3. Fracture energy evaluated according to [17]

Set	G_F
	[N/m]
C	89.7

Steel. For the reinforcing steel the regular hot-rolled steel was used. The yield strength, the ultimate strength, the Young's modulus and the yield strain are shown in Table 4.

Table 4. Steel properties

ϕ	f_y	f_u	ε_u	E
mm	[N/mm²]	[N/mm²]	[%]	[N/mm²]
6	578.3	638.6	16.3	200366
12	580.0	631.5	23.0	198250

Test - Setup

The beams were tested in a three-point-bending structural system which is shown in Fig. 12. The load was applied with a load jack in the middle of the beam over a loading plate. Cylindrical steel bearings assured free lateral translation and rotation of the supporting load plate (Fig. 13). The load was applied by controlling the mid-span displacement using a servo-hydraulic Schenck machine. The displacement rate was $v = 0.002$ mm/s according to [18]. After the formation of the first bending crack the loading rate was increased to $v = 0.01$ mm/s.

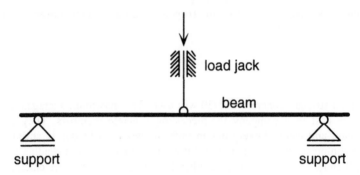

Fig. 12. Structural system.

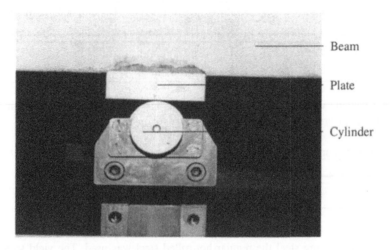

Fig. 13. Support.

During the experiments the forces were measured by load cells which were installed in the loading jack. The displacements of the hydraulic cylinders were recorded as well. The crack opening and the mid-span-deflection were measured with electric resistance transducers. The transducers had a measuring length of 75 mm. The measuring distance of the crack evaluating transducers were $\ell_b = 100$ mm (Fig. 14a). There were two transducers for the mid-span-deflection in case of a rotation of the beam (Fig. 14b). Every signal was monitored and recorded with a frequency of 2 Hz by an amplifier from HBM.

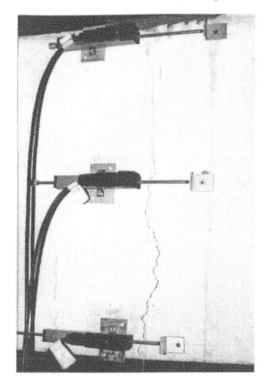

Fig. 14a. Crack opening.   Fig. 14b. Mid-span-deflection.

Results

General. For every test set (A,B,C) two identical tests were carried out. The load-strain curves measured in the experiments are shown in Figs. 15a, 16, and 17a. The strains were calculated by dividing the displacements measured with LVDT 1 by the measuring distance $\ell_b = 100$ mm (Fig. 15a). The transducer LVDT 1 was fixed exactly at the level of the reinforcement. The computed yield load and ultimate load of the system are shown in Figs. 15a, 16, and 17a as well. These loads were calculated according to the cross sectional analyses, assuming a bilinear stress-strain curve for steel and a parabolic-rectangular stress-strain curve for concrete. The material properties were taken from the experiments. The maximum concrete strain was assumed to $\varepsilon_{cu} = -3,3 \cdot 10^{-3}$. In each diagram the load-strain curve for the corresponding steel is shown as well. After the yield stress is reached a linear behaviour for steel is assumed.

Set A (h = 0.125 m). The first crack occurred at the total load of 10.2 kN. The measured concrete strain was 0.019 %. This is corresponding to a tensile strength of 2.53 MPa, evaluated according to linear elastic beam theory. In total three cracks formed (Fig. 15b). They were located and spaced exactly at the stirrups positions. The measured ultimate load was $F_U = 27$ kN. After the peak was reached one of the steal bar ruptured and the load decreased suddenly to 13.7 kN. Later on the load recovered and reached 17.6 kN (Fig. 15a). Subsequently the second reinforcement bar failed and the test was stopped.

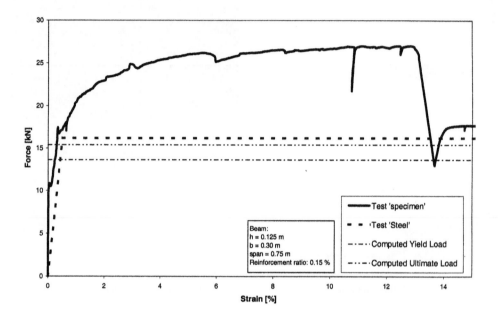

Fig. 15a. Force-strain curve of beam A (h = 0.125 m).

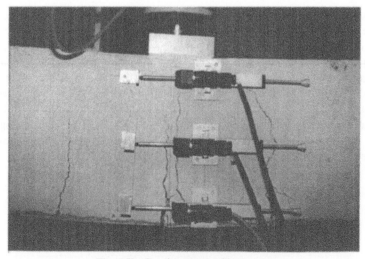

Fig. 15b. Crack pattern of beam A.

Set B (h = 0.25 m). By testing beam B the first crack opened at the load of 19.5 kN. The measured concrete strain was 0.02%. This is corresponding to a tensile strength of 2.5 MPa. In total three cracks were observed. The locations of the cracks corresponded to the locations of the stirrups as it was seen in test A. The maximum load was F_U = 47.5 kN. After a smooth softening branch the load decreased suddenly to 33.4 kN. Like in set A ,the reinforcing bars failed one by one, showing a steep softening branch. After each rupture, the load recovered until the last bar failed (Fig. 16a). The necking of the reinforcement could be seen in Fig. 16b.

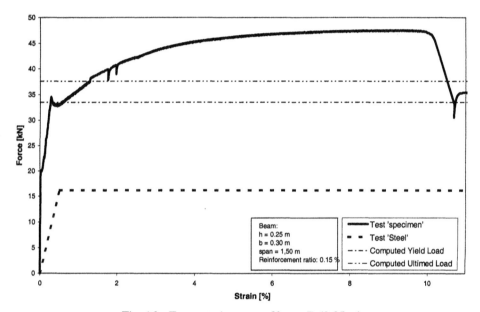

Fig. 16a. Force-strain curve of beam B (0.25 m).

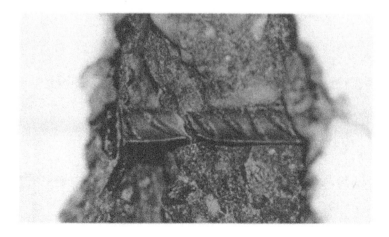

Fig. 16b. Rupture of steel.

Set C (h = 0.50 m). Beam C cracked at a load of 54.4 kN and the measured concrete strain was 0.003 %. The calculated tensile strength is in total 3.6 MPa. In total eight cracks were observed (Fig. 17b). As in the above tests, the crack localisation coincides with the stirrups position. The maximum load was F_U = 89.0 kN (Fig. 17a). After a relatively long softening branch the reinforcing steel bars failed all at once and the load dropped to zero.

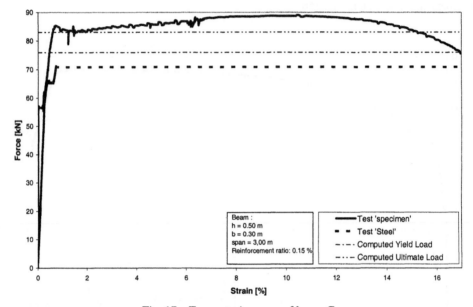

Fig. 17a. Force-strain curve of beam C.

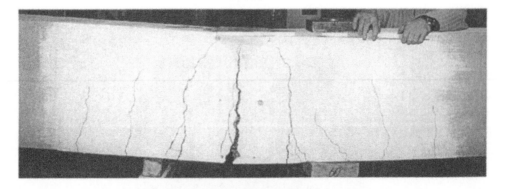

Fig. 17b. Crack spacing.

DISCUSSION OF THE NUMERICAL AND EXPERIMENTAL RESULTS

The main aim of the minimum reinforcement ratio is to prevent the brittle failure of reinforced concrete beams and to make the redistribution of internal forces possible. The numerical and experimental results show that small beams with low reinforcement ratio exhibit a ductile response with significant contribution of concrete to the peak resistance (see Fig. 15a and 16). This is due to the size effect on the bending strength of concrete as well as to the interaction between reinforcement and concrete. Since these beams exhibit significant ductility it seems that there is no need to increase the amount of minimum reinforcement in small beams which are relatively ductile anyway. Therefore, instead to use Eq. (1) proposed by [3], Eq. (3) seems to be more appropriate for small RC beams i.e. the minimum reinforcement should not depend on the size of the beam.

In contrast to the small beams, the numerical results for large beams indicate that the energy criterion, Eq. (5), is relevant for the minimum reinforcement requirement, i.e. when the beam size increases the minimum reinforcement ratio increases as well. However, this seems to strongly depend on the amount and type of distributed reinforcement, type of the main reinforcement as well as on the bond capacity between reinforced bar and concrete. In the present numerical study the concrete was assumed to be rather brittle and the bond capacity relatively decreased with the increase of the beam size. Consequently, the minimum reinforcement ratio increased with the increase of the beam size. However, when introducing distributed reinforcement or using more ductile concrete with good bond properties the minimum reinforcement ratio seems to be not size dependent. Presently, there exists only experiments for the beam depth of $h = 500$ mm with $\rho = 0.15\%$ provided with stirrups according to [6]. The test on these beams did not confirm the numerical prediction since, apart from the bond properties, the brittleness of concrete was in the analysis of about three times higher than in the tests. Therefore experiments on large beams are needed and they are currently in progress.

Due to the high brittleness the energy aspect of the minimum reinforcement ratio is especially important for beams made of high strength concrete. Dynamical effects due to the acceleration of the beam when the crack propagates are important as well. These effects may be so strong that the minimum reinforcement according to Eq. (3) could not be able to prevent an unstable crack propagation. In the future these aspects of the minimum reinforcement requirements should be investigated in more detail.

CONCLUSIONS

- The numerical and test results for small beams with low reinforcement ratio show a ductile behavior with a significant contribution of concrete to the peak load. With decreasing the beam size the ductility and contribution of the concrete to the resistance increases.
- For small beams the lowest possible reinforcement ratio could be based on the stress equilibrium criteria and it should not depend on the beam size. For these beams contribution of the concrete to the ultimate resistance and ductility should not be neglected. This showed both, tests and numerical results.
- According to the numerical results, after reaching a critical beam size the minimum reinforcement ratio increases with increase of the size. The increase of the minimum reinforcement is governed by the energy equilibrium after the first bending crack appears.
- The amount of the minimum reinforcement as well as the transitional beam size (small to large) depends on: (1) Brittleness of the concrete, (2) concrete-steel bond relationship and (3) type and amount of the distributed reinforcement. The highest increase of the minimum reinforcement ratio

and the smallest transitional size should be expected for brittle concrete (high strength concrete) with no distributed reinforcement. To proof or disapprove this further experiments on large beams made of normal and high strength concrete are in progress.

- Beams with a depth $h > 1.00$ m designed according to [16] must have a distributed reinforcement. The numerical results for these beams show a ductile behavior and the minimum reinforcement requirements given in the [16] seems to be correct.
- There is a clear indication that the importance of the distributed reinforcement becomes more significant in larger beams. Apart from other effects such as temperature, shrinkage, etc., from the fracture point of view the distributed reinforcement is less important in small beams.

REFERENCES

1.   ACI Committee 318, (1989). *ACI* Building Code Requirements for Reinforced Concrete, American Concrete institute, Detroit, MI. pp.353.
2.   Bazant, Z.P. (1987). "Snapback instability at crack ligament tearing and its implication for fracture micromechanics,"*Cement and Concrete Research* **17**, pp. 951-967.
3.   Bosco, C., and Carpinteri, A. (1992). "Fracture mechanics evaluation on minimum reinforcement in concrete structures," *Application of Fracture Mechanics to Reinforced Concrete*, Ed. by A. Carpinteri, Elsevier Applied Science, Torino, Italy, pp. 347-377.
4.   Bigaj, A., and Walraven, J.C. (1993). "Size effect on the rotational capacity of plastic hinges in reinforced concrete beams,"*CEB*-Bulletin 218.
5.   Carpinteri, A. (1981). "A fracture mechanics model for reinforced concrete collapse," *IABSE* Colloquium on Advanced Mechanics of Reinforced Concrete, Delft, pp. 17-30.
6.   CEB, (1990). *CEB-FIP Model Code* -- Final Draft, Comitee Euro-International du Beton, Paris.
7.   Hawkins, N.M. (1992). "Minimum reinforcement requirements for concrete flectural members," *Application of Fracture Mechanics to Reinforced Concrete*, Ed. by A. Carpinteri, Elsevier Applied Science, Torino, Italy, pp. 379-412.
8.   Hillerborg, A. (1989). "Fracture mechanics and the concrete codes," *Fracture Mechanics: Applications to Concrete*, ACI-SP118, Ed. V. Li and Z.P.Bazant, pp. 157- 70.
9.   Hu, C. (1986). "Ductility of concrete slabs reinforced with welded wire fabrice," M.S.C.E. thesis, University of Washington, Seattle, WA, USA.
10. Jenq, Y.S., and Shah, S.P. (1989). "Shear resistance of reinforced beams -- a fracture mechanics approach," *Fracture Mechanics: Application to Concrete*, V.C. Li, and Z.P. Bazant, eds., SP-118, ACI, Detroit, pp. 237-258.
11. Karihaloo, B. L. (1992). "Failure modes of longitudinally reinforced beams," *Application of Fracture Mechanics to Reinforced Concrete*, Ed. by A. Carpinteri, Elsevier Applied Science, Torino, Italy, pp. 523-546.
12. Karihaloo, B. L., Carpinteri, A. and Elices, M. (1993). "Fracture mechanics of cement mortar and plain concrete,"*Advanced Cement based Materials*, **1**, pp.92-105.
13. Ozbolt, J. (1995). "Maßstabseffekt und Duktilität von Beton- und Stahlbeton Konstruktionen." Postdoctoral Thesis, Universität Stuttgart.
14. Ozbolt, J., Y. Li and Kozar, I. (1996). "Microplane model for concrete – mixed approach" Submitted to *IJSS*.
15. Ozbolt, J., and Bazant, Z.P. (1996). "Numerical smeared fracture analysis: Nonlocal microcrack interaction approach." *IJNME*, **39**(4), pp. 635-661.
16. Eurocode 2 (1992) Planung von Stahlbeton- und Spannbetontragwerken, Teil 1: Grundlagen und Anwendungsregeln für den Hochbau.

17. RILEM Draft Recommendations (1985) Determination of the fracture energy of mortar and concrete by means of three point bend test on notched beams, Materials and Structures, **18**, pp 285-290.

18. DIN EN 12359 (1996) Bestimmung der Biegezugfestigkeit von Körpern.

Author Index

Printed and bound by CPI Group (UK) Ltd, Croydon, CR0 4YY

03/10/2024

01040320-0010